1700 1750 18

Römer

tholinus

ygens Euler

ss

Fraunhofer

Leibnitz Lambert Döppler

Dollond

Newton Young

ooke

rdies

Fresnel

Malus

Arago

ontanari Melloni

Boscovich

1700 1750 1800 1850

The Nature of Light

FRONTISPIECE ILLUSTRATION

Top Diffraction pattern seen when observing a point source of white light through a mesh with approximately 25 squares per square millimetre. An observation of this type was carried out by Grimaldi.

Lower left Diffraction pattern seen on a white screen, produced by a narrow slit illuminated by a point source of white light.

Lower right Diffraction pattern seen on a white screen produced by a pinhole illuminated by a point source of white light. The diameter of the hole is much smaller than that of the bright spot shown in the picture. This observation was described by Grimaldi and this figure is now known as 'Airy's disc' or 'centric'.

The Nature of Light

AN HISTORICAL SURVEY

Vasco Ronchi

TRANSLATED BY V. BAROCAS

HARVARD UNIVERSITY PRESS
CAMBRIDGE, MASSACHUSETTS
1970

Originally published in Italian as
Storia della Luce, 1939
English version with new material by the author

First published 1970

1970 Importation

SBN 674-60526-8

Printed in Great Britain

Translator's Foreword

THE UNDERSTANDING of the nature of light has progressed very rapidly during this century. The foundations of our knowledge of this subject, however, go back many centuries, almost to the dawn of civilization.

When the ancients began to try to understand their environment they naturally gave considerable thought to 'light' which pervaded their universe, and they began to advance theories which could explain its nature. Soon there came the realization that light could not be considered objectively and that the individual's reaction to it played a very important part. As a result the study of light itself became involved with the anatomy and physiology of the eye as well as with psychology.

The ideas of the early philosophers were very confused, and as they appeared to bear very little relation to the physical studies of light of later centuries, most of them were soon forgotten by physicists. Indeed in many cases even important theories, in their own day, were not fully appreciated by students of optics and were completely ignored, thereby considerably delaying the interpretation of some fundamental phenomena. There is no doubt however, that our present knowledge about the nature of light is in no small measure due to the pioneering efforts of the early thinkers.

This book traces the development of these ideas from the golden age of Greece to the nineteenth century when so much was discovered about light and when a coherent picture began gradually to emerge. The reader must not expect to find here detailed descriptions of optical instruments or mathematical discussions. The aim of the book is to give in a logical sequence the development of ideas which led to our present knowledge about the nature of light, and in it we find a broad survey of ideas related to the meaning and to the essence of light. Thus we are able to follow the vicissitudes of many theories, some of which survived although in a modified form, and some which had a very limited life. Through the pages of this book the complexity of the question becomes apparent and we acquire an insight into the long journey that the human mind had to make before reaching a satisfactory understanding of some of the more complex phenomena of light.

The author of *The Nature of Light*, Professor Ronchi, is an acknowledged authority in the field of optics. He has published many works on the subject and the 'Ronchi test' is widely known in the world of optics. In 1927 he founded the *Istituto Nazionale di Ottica* at Arcetri, near Florence, and he became its director, a position which he still holds today.

It is natural that an enquiring mind such as that of Professor Ronchi should in due course turn to the history of science and in particular to the question of the nature of light. As a result of this he wrote his *Storia della Luce* in 1939, which was followed in 1952 by a second edition. The book made a notable contribution to the subject and was welcomed by many of his international colleagues. In 1956 a French translation appeared in Paris. The present English version arises from our desire to make this book available to the English-speaking world.

The English translation is based mainly on the second Italian edition. The original work was revised in conjunction with the author, with the addition of new material and new illustrations.

We hope that this translation will convey the inspiration of the author and show his valuable contribution to the history of science.

The Jeremiah Horrocks and Wilfred Hall Observatories, Preston. — *V. Barocas*
July 1970

Contents

List of Plates

COLOUR PLATES

BLACK AND WHITE PLATES

CHAPTER ONE

Optics in the Graeco-Roman age

'*And God said Let there be light: and there was light.*' (*Gen. I v.* 3).

GOD began His immense work of Creation by giving to heaven and earth this wonderful gift of light.

Whatever may be the philosophy and the faith of anyone who considers this fact, the deep meaning contained in the first verses of *Genesis* cannot be doubted. Light is of vital importance for the world. Light is the basis of life and order as opposed to darkness, chaos, disorder and death. Without light the earth was 'without form and void'.[1] Even today when we refer to the birth of a child we still say 'he saw the light of day'.

The first verses of *Genesis* are of very great importance because they give to light an absolute precedence of Creation. Precedence over the sun[2] which was created on the fourth day and over man who was created on the last day. In these circumstances and in these words a theory on the nature of light is already implicit. This theory attributes to light an existence of its own, independent of its source and of its receiver.

The books of the Old Testament originated in an Eastern civilization which spread into Western Europe with the advent of Christianity. It would certainly be of great interest to examine and study the old texts of the various centres of ancient civilizations because in these texts we would doubtless find the ideas and the theories formulated by the sages of the time, with reference to the nature of light. Studies of this period have been possible only for one centre of civilization, namely ancient Greece, which fortunately is a centre of extreme importance.

It is well known that Greek philosophy had its golden age in the fifth and fourth centuries B.C. This is a period that even today, twenty-four centuries later, is a source of wonder and of admiration not only for historians. The names of the philosophers and sages of that period are even today held in great esteem as a symbol of

[1] See *Genesis* I vv. 1–5

[2] The verses in *Genesis* which refer to the creation of the sun (vv. 14–19) are of great interest

knowledge and intelligence which are almost divine. The foundations of the philosophical sciences of those days still form the basis of our modern science. Time and progress have not destroyed them, on the contrary, later generations have built upon that knowledge. It was inevitable that the Greek sages should turn their attention to light and should attempt to formulate theories about it. Indeed their work deserves special study. From the seventeenth century onwards, attempts have been made to discredit this body of knowledge. Nevertheless we must admit that the critics themselves were confused, perhaps because they were mesmerized by the temporary successes of the new 'natural philosophy'. In saying this we are not asserting that all the ideas of the Greek philosophers are unassailable. Much of what has come down to us can withstand the most strict criticism, provided such criticism is honest and without prejudice. The majority of these ideas deserve to be considered as strictly logical reasoning, carried out not only with common sense, but even with noticeable acumen. If we cannot appreciate fully both the physical and philosophical theories of light developed by the Greek philosophers, this is probably due to the difficulty of interpreting the true significance of the ancient texts. Unfortunately most of the information we have is second-hand and therefore affected by an interpretation that may well have distorted it. Very probably this has happened.

The direct knowledge that has come down through the centuries to us is only fragmentary. Philologists are often faced with real difficulties when they have to interpret terms that have a significance that is both wide and indeterminate. It is almost certain that the original authors used these terms with a much more definite significance than that given to the words when interpreted by modern etymology and philology; this is the only interpretation available to the modern translators and commentators. These people are in a worse position than that facing the translator of a scientific text-book, even a modern one, who has no technical knowledge of the subject dealt with by the book. So it happens that when we try to interpret the Greek texts, we are faced with terms that are so flexible that their meaning may not reveal the real concept but depend on the mental approach of the translator himself.

The Greek philosophers do not appear to have taken upon themselves the task of determining the nature of light. What interested them most was to explain the mechanism of vision. In those days the main goal of thinkers was to learn to understand man, his functions and his faculties. Vision was one of the important faculties of man, and hence the answer to the question 'how do we see?' became fundamental. Every physical entity exists because it produces effects. At that time the only known effect of light was vision, and it was natural therefore, that the study of light should begin from this point.

If we start our study from the fifth century B.C., which was the highest peak of the Greek age, we find that already there existed schools of thought engaged in the discovery of the mechanism of vision. This indicates that earlier there must have been a period of preparation which we shall not attempt to investigate even summarily since we have so very few ancient texts on this subject. The distinction between the various schools of thought is made possible by the great distance of time, when we are able to ignore details and concentrate only on the main outline. It is probable, however, that during that century there were many thinkers engaged in discussions and debates on this subject which they considered to be of extreme importance. These discussions must have been very varied, on account of the many philosophical systems then existing, which were bound to affect the approach to the problem by introducing new theories, hypotheses and ideas.

In the fifth century B.C. there was, no doubt, an unceasing attempt to determine the link between the eye and the object seen. This link is what interests us in particular and is what many today call 'light'. The ancient thinkers did not however, reach immediately either the proof or the hypothesis of the existence of this something objective called 'light', which linked the object seen with the seeing eye. If nowadays, we use this expression without reservation and without even thinking, as if it were an evident reality, it is because of the studies and of the discussions which have gone on for over two thousand years. In the fifth century B.C. this type of link between the eye and the object seen was rightly considered as a possible solution, but not the only one. The link could be thought to be due to 'something' (*quid*) which was emitted by the eye and travelled towards the object, or to 'something' which travelled towards the eye from the object, or finally to the co-existence of both these 'somethings', travelling in opposite directions. Even with this, all the possible solutions were not exhausted, since some other type of link could be imagined, without any motion in either direction, but merely a modification of the medium interposed between the eye and the object seen. In the discussions which were taking place at the time, all these ideas were put forward.

The first idea, that of the emission from the eye travelling to the object, was accepted by the Pythagoreans. The Atomists of the school of Democritus naturally were in favour of the theory of an emission from the object travelling towards the eye. Empedocles was among the first to support the idea of a combination of two fluxes. It was only in the following century that a fourth hypothesis was added. This is found in the works of Aristotle, perhaps because by then there was already some dissatisfaction with the theories which had been advanced. We can say from the outset that none of these theories was ever fully accepted. Indeed even many centuries after, we find followers of one or other of the theories, though there was a tendency to support the more extreme views.

The information we have available is so widely distributed over the centuries that it does not permit us to follow step by step the evolution of these ideas. In order to avoid any confusion which would inevitably follow if we were to examine the texts in a strict chronological order, we shall try to study each theory independently and follow its development even if, occasionally, this method will take us back into earlier times.

The early formulation of a concept concerning the nature of light is attributed to Empedocles of Agrigentum (*c.* 500–*c.* 430 B.C.). According to him, light was an emanation originating from all bodies and reaching the eye. He considered the structure of matter and reached the conclusion that matter must be porous, since it shows pores, namely tiny holes between the solid elementary particles. Generally speaking he assumed that sensations were felt when certain particles emitted by various bodies, entered the pores of the sensitive organs and these pores were arranged in such a way as to let through the particles. In the particular case of the sense of sight, light would be the emanation of an external elementary fire which would reach the eye, across space, in the same way as sound reached the ear and smell reached the nose. We must note here that such an expression as 'external elementary fire' must not be interpreted in the meaning that we have learnt to give to these words in our times. Together with this external action, Empedocles claims that it is necessary that there should be also an action from the interior of the individual to the exterior:

> Thus black and white and every other colour will appear to us as produced by the encounter of our eyes with something which moves in the direction of the eyes; and that any particular colour which we see is neither the object which comes towards the eye nor the eye which is met, but rather something which is produced between them . . . It is necessary that if a certain thing becomes sweet or bitter, or has another taste, it is possible but to sense nothing is not possible. Similarly, it is necessary that if a certain thing becomes sweet or bitter, or has another taste, it must become so for someone, since if it is possible to become sweet, it is not possible to become sweet for no one.[3]

On the basis of this reasoning and of the principle of attraction of like to like, Empedocles sees the possibility of some reciprocity, and considers that there must be an emission of rays from the interior to the exterior of the eye, simultaneously with the emission directed from the exterior to the interior. He therefore, considers the possibility of the existence of two fluxes: one external, existing by itself, objective and of a corpuscular nature, carrying with it the shape and colours of the objects. This however, was not sufficient to explain what one could see. In order to explain what was already known about vision it was necessary to consider the existence also of rays which were emitted by the eyes.

[3] See F. Enriques and G. de Santillana: *Storia del pensiero scientifico*, Vol. 1 'Il Mondo Antico', Bologna, Zanichelli, 1932, p. 165

Such a schematic and concise interpretation of Empedocles' ideas is very debatable because his reasoning is less precise and more complex and his own expression of these ideas is extremely difficult to translate. Let us not forget that we are dealing with that period of Greek philosophy when an elementary significance was attributed to the words water, fire, earth and air which is certainly not there today. Thus Empedocles states that the interior of the eye is water and fire, and the exterior world is earth and air. Later, however, he mentions eyes with an internal fire and eyes with an external fire, and on this he tries to base an explanation of why some animals see better in daylight while others see better at night.

We do not consider it necessary for our purpose to pursue a deeper critical study of what Empedocles was trying to say. This study would require a careful examination of all the various texts referring to his work. For our purpose it suffices to note the fundamental idea, namely the attempt to combine an intervention of external material particles with the internal action of the eye by means of a 'fire', a word which could be interpreted as meaning spirit, soul or some other entity even less well-defined.

Leucippus of Miletus, whose name is often linked with that of Democritus, was a contemporary of Empedocles. The following type of reasoning is generally attributed to Leucippus:

> ... Every modification, either produced or received, is due to a contact; all our perceptions are due to the sense of touch and all our senses are but variations of the sense of touch. As a consequence, since our soul does not come out to touch the objects of the outside world, it is necessary that these objects themselves come to touch our soul by passing through our senses. Now we do not actually see the objects coming nearer to us when we perceive them, therefore, they must send to our soul 'something' which represents them, some image, ειδολα, some kind of shadow or some material simulacrum which envelopes the bodies, quivers on the surface and can detach itself from them in order to bring to our soul the shape, the colours and all the other qualities of the bodies from which they emanate.[4]

[4] See J. Trouessart: *Recherches sur quelques Phénomenes de la Vision.* Brest, Anner, 1854, p. 15

Democritus of Abdera (*c.* 460–*c.* 360 B.C.) put forward ideas which appeared less definite, at least from what we learn from Theophrastus. In the ideas attributed to Democritus we catch a glimpse of the existence of an idea of particular interest. He considered that 'the air interposed between the eye and the object, receives an imprint as a result of the pressure exerted upon it both by the eye and by the object'. This concept however, remained mixed with other similar ideas; that from all objects there was a continuous emission, and that the substance emitted being solid and differing in colour from the eye, imprinted itself on the eyes which were moist. Democritus must have noticed that small images of objects which were in front of the eyes could be seen on the cornea. He then studied in particular the liquid of the eye and the sponginess and the permeability of its various membranes. Here for the first time we meet the idea of a 'transmission medium' if, however, the interpretation of the original text has not been completely misunderstood.

Even before Democritus, however, the Pythagoreans had advanced other ideas which increased the great variety of theories existing at the time, concerning the great mystery of vision. According to the writings of Apuleius, Archytas of Taranto (*c.* 430–*c.* 305 B.C.) who was a contemporary and a friend of Plato and a worthy exponent of the Pythagorean school, believed that vision was due exclusively to an invisible 'fire' which came out from the eyes to touch the objects and to reveal their shape and colour. It would appear that gradually the discussion of these theories led to the formation of different schools of thought. On one hand the Atomists adhered to their views which were similar to those attributed to Leucippus and on the other hand the Pythagoreans denied every external influence and maintained that the eye had the ability to see objects by emitting a 'something' suitable for this purpose. This particular idea of the Pythagoreans had surprising results in the following centuries. In fact it was against this particular idea that all the criticism of the physicists of the last few centuries was unleashed. They used this to demonstrate how little physics was known in the Greek period of the history of science. We, on the other hand, cannot ignore the evolution of this theory because it had a very great influence on the development of optics. Once again unfortunately, the fragmentary information which has reached us, does not allow us to follow its evolution step by step. However, it is true that we still find these ideas very much alive over a period of seven centuries in the minds of men like Euclid, Hipparchus, Ptolemy and Damianus. The difficulty is to be able to define what exactly Archytas meant by the invisible 'fire' which flows from the eyes. We must remember that expressions of this type have been used widely, without any scientific meaning, by poets throughout the ages, not excluding our contemporaries. In novels and fairy tales we often find reference to emanations from the eyes of witches, of wild animals, of snakes, of lovers or of followers of black magic. Is it possible that in all this there is absolutely no truth?

We propose to carry out our investigations without bias as far as we can; we must not exclude however, ideas which have been expressed by men of genius and which have reached us through the centuries, even if they do not agree with modern science. In passing, we may note as a matter of interest, the variety of conclusions which can be reached by the interpretation of ancient ideas according to whether it is made in good or bad faith. A hostile critic, from the standpoint of modern physics, may consider the ideas attributed to Archytas to be completely false and ridiculous. A more kindly inclined critic may think that he recognizes in these ideas the mechanism of vision by means of the projection to the exterior of the retinal images, which is a mechanism nowadays generally accepted. In the field of physics the favourable critic may see in Archytas the forerunner of comparatively modern physicists, who have collected in one group both visual and thermal radiations. Archytas has in fact compared the 'fire' to the agent of vision, namely to 'light'.

Many discussions and debates must have taken place on these theories, since the original texts which report them at various intervals of time, show a progress and an evolution of concepts which tend to become gradually clearer and better defined. That all the philosophers of the time showed great interest in this particular question and contributed widely towards a better understanding of it, is shown by the many references we find in the various dialogues of Plato (*c.* 428–*c.* 347 B.C.).

In *Meno* we find again the definition of light and of colour which had been put forward by Gorgias of Lentini (Sicily, born 480 B.C.) following the ideas of Empedocles: 'colour is the emanation from the body which is perceived, corresponding to vision'.

In *Timaeus* we find the following:

> . . . They (the gods) contrived that all such fire as had the property not of burning but of giving a mild light should form a body akin to the light of every day. For they caused the pure fire within us, which is akin to that of day, to flow through the eyes in a smooth and dense stream; and they compressed the whole substance, and especially the centre, of the eyes, so that they occluded all other fire that was coarser and allowed only this pure kind of fire to filter through. So whenever the stream of vision is surrounded by mid-day light, it flows out like unto like, and coalescing therewith it forms one kindred substance along the path of the eye's vision, wheresoever the fire which streams from within collides with an obstructing object without. And this substance, having all become similar in its properties because of its similar nature, distributes the motions of every object it touches, or whereby it is touched, throughout all the body even unto the soul, and brings about that sensation which we now term 'seeing'. But when the kindred fire vanishes into night, the inner fire is cut off; for when it issues forth into what is dissimilar it becomes altered in itself and is quenched, seeing that it is no longer of like nature with the adjoining air, since that air is devoid of fire. Wherefore it leaves off seeing. . . .[5]

In *Theaetetus* we also find a very interesting passage:

> SOCRATES: Then just take a look round and make sure that none of the uninitiate overhears us. I mean by the uninitiate the people who believe that nothing is real save what they can grasp with their hands and do not admit that actions or processes or anything invisible can count as real.
> THEAETETUS: They sound like a very hard and repellent sort of people.
> SOCRATES: It is true, they are remarkably crude. The others, into whose

[5] See *Timaeus* 45C translated by the Rev. R. G. Bury-Heinemann, London, 1929, p. 101. See also U. Forti: *Le concezioni della luce da Democrito a Cartesio.* 'Periodico di Matematiche', 1928, p. 92

secrets I am going to initiate you, are much more refined and subtle. Their first principle . . . is that the universe really is motion and nothing else. And there are two kinds of motion. Of each kind there are any number of instances, but they differ in that the one kind has the power of acting, the other of being acted upon. From the intercourse and friction of these with one another arise offspring, endless in number, but in pairs of twins. One of each pair is something perceived, the other a perception, whose birth always coincides with that of the thing perceived. Now, for the perceptions we have names like 'seeing', 'hearing', 'smelling', 'feeling cold', 'feeling hot', and again pleasures and pains and desires and fears, as they are called, and so on. There are any number that are nameless, though names have been found for a whole multitude. On the other side, the brood of things perceived always comes to birth at the same moment with one or another of these—with instances of seeing, colours of corresponding variety; with instances of hearing, sounds in the same way; and with all the other perceptions, the other things perceived that are akin to them. . . . The point is that all these things are, as we were saying, in motion; but there is a quickness or slowness in their motion. The slow sort has its motion without change of place and with respect to what comes within range of it, and that is how it generates offspring; but the offspring generated are quicker, inasmuch as they move from place to place and their motion consists in change of place. As soon, then, as an eye and something else whose structure is adjusted to the eye come within range and give birth to the whiteness together with its cognate perception—things that would never have come into existence if either of the two had approached anything else—then it is that, as the vision from the eyes and the whiteness from the thing that joins in giving birth to the colour pass in the space between, the eye becomes filled with vision and now sees, and becomes, not vision, but a seeing eye; while the other parent of the colour is saturated with whiteness and becomes, on its side, not whiteness, but a white thing, be it stock or stone or whatever else may chance to be so coloured.

And so, too, we must think in the same way of the rest—'hard', 'hot' and all of them—that no one of them has any being just by itself . . . but that it is in their intercourse with one another that all arise in all their variety as a result of their motion. . . .[6]

[6] See F. M. Cornford. *Plato's Theory of Knowledge*. Routledge and Kegan Paul Ltd, London, 1935, p. 46

A very narrow and unsympathetic interpretation of these extracts from Plato's dialogues and from other similar texts have led many to suggest almost in a derisory manner that the theory consisted in accepting the existence of two types of rays. One external, travelling from the object to the eye and the other from the eye directed towards the object. The meeting of these two emanations half-way gave birth to vision. When these are expressed in these terms they certainly do not deserve much consideration. We do not believe, however, that Plato meant this. Apart from the fact that he concentrated more on the question of perception in general than on the visual perception in particular, it is obvious that he insisted above all on the psychological aspect of vision. Plato felt the need of an external agent (the brightness which emanated from the object and reached the eye, namely light) and of an internal agent (namely the visual fire) emanating from the eye in order to give life and substance to the object seen, which was what it appeared as a result of the whole process.

Making allowances for the considerable difficulties required in the interpretation of the words of the author, we cannot deny that the concept, as we see it, is basically rational. Moreover, we must remember that it is very difficult to speak of a real theory belonging to Plato since the *Dialogues* reflect the many opinions which were held at the time. In fact there is no real agreement between the definition of light attributed to Gorgias in the *Meno* and the idea of light as motion found in the *Theaetetus*.

The suggestion that light was of a mechanical nature rather than material, outlined by Democritus and now supported by Plato, was further developed in the works of Aristotle of Stagira (384–322 B.C.). His writings are not very clear and, as usual, interpreters have read into them what they wished to find. It is a fact that although in some passages he appeared to support a material theory of light, in the greater part of his work in which he dealt with vision[7] he developed and defined the idea barely outlined by his predecessors, namely that light was a phenomenon of a mechanical

[7] See: *On the Soul* (*De Anima*) Book II Chap. VII: *On the Senses* Chap. II and III: *On the Parts of Animals* Book V Chap. I

nature. Following the custom of his times, he did not mention any of his predecessors who had advanced ideas of a similar nature, but only those writers of opposing views against whom he had plenty to say.

Aristotle did not accept the hypothesis of the emission of light from the eye which, in his opinion, came from Empedocles. He criticised this theory in the following words in Chapter II of *On the Senses*:

> If the eye was made of fire as Empedocles says, and as it is suggested in *Timaeus*, if vision were produced by means of a fire emitted by the eye, like the light emitted by a lantern, why then are we not able to see in the dark? To say that this light is extinguished as it spreads out in the darkness, as it is stated in *Timaeus*, is completely meaningless. How can light be extinguished? Heat and dryness are extinguished by cold and dampness, and we have an example of this in the fire and flames which come from glowing charcoal. Neither heat nor dryness however, appear to be attributes of light. If they existed in it and were not visible because of their inactivity, it would follow that on a rainy day, light would be extinguished and that on a freezing day we would have darkness, since this is the fate which flames and incandescent bodies suffer. Nothing of the sort happens.

Aristotle also did not agree with Democritus, who referred to the impression of images on the pupil, perhaps because he was misled by the images seen by reflection on the cornea. This process was clearly understood by Aristotle although he did not express it in such modern terms.

He also declared himself against the theory contained in the *Theaetetus*:

> It is completely absurd to maintain that we can see because of 'something' which is emitted by the eye and that this 'something' may reach as far as the stars or until it meets another 'something' coming towards it, as other philosophers have suggested. If anything it would be preferable to consider that this union should occur within the eye itself. However, it would be foolish to accept even this since, after all, we do not understand the meaning of this union of light with light and we do not even understand how it could take place.

As for the supporters of the theory of the emission of simulacra he disposed of them with a few words in Chapter III of *On the Senses* at the same time demolishing the hypothesis of 'touch' attributed to Leucippus.

> To say, as the Ancients did, that colours are emissions and that this is how we see, is absurd. First of all they should have proved that all our perceptions are due to the sense of touch. . . . [He concluded] Once for all, it is preferable to accept that perception arises from a movement, produced by the body we perceive, in the interposed medium, rather than to consider it to be due to a direct contact or to an emission.

This criticism though summary, is clear. Many objections could easily be raised to all the three theories discussed. Probably, even in those days, the general opinion must have been that the truth was yet to be found. There was not therefore, a real need for criticism, there was rather a need for constructing new theories, after having demolished completely the disjointed arguments of the earlier thinkers. The theories Aristotle put forward are not as clear as his criticism. The principal interpreters of his ideas, including the ancient interpreters, suggested that Aristotle was thinking in terms of a movement travelling between the object and the eye and modifying the state of diaphanous bodies. A diaphanous body in the dark was in a potential state; it was potentially diaphanous. The same body was said to be in the light when it was actually diaphanous.

In Book II, Chapter VII of *De Anima*, Aristotle expressed himself thus on this question:

> We have explained what is diaphanous and what is light and how the latter is neither fire nor generally, a concrete element, nor an emanation of any body (since in this case it would be a concrete element) but it is the result of the existence of a fire or something similar, in contact with something diaphanous, since it is not possible that two bodies should exist simultaneously in one place. . . .

There is little doubt that Aristotle was determined to exclude any emission which, according to him, would be contrary to the impenetrability of bodies. We must admit that it is not very easy to understand the meaning attributed to the words he wrote about his ideas. Commentators would have to be very biased in order to see in the words 'result of the existence of a fire' the expression of a present action, namely a modification of the medium due to the source of fire. The more favourable interpreters linking this passage with some others, see in Aristotle nothing less than a forerunner of the modern school of the wave theory of light. Perhaps those who interpret these writings in this way, are not very far from the truth. After all, having rejected the existence of an emanation from the eye, the emanation from the bodies and a combined emanation from both, nothing remained but to have recourse to a motion or to an alteration of the medium. Unfortunately it looks as if Aristotle lacked words to express his thoughts, if indeed these were his thoughts. In other passages, trying to explain his thoughts, he went as far as to say that light was like the soul of a diaphanous body, its very life, while the same diaphanous body when in darkness was as if dead. As a confirmation of this thought, he added that if a total vacuum existed around the eye, that is to say if there were an absence of all media (including the ether), vision would be impossible.

The ideas of Aristotle, like those expressed in Plato's *Dialogues*, do not seem to have been accepted or assimilated by his contemporaries. Examination of documents relating to later periods show that the theories of the Pythagoreans and of the Atomists fared much better than the more complex theories which abounded in the Greek age.

The work of Euclid, who was a pupil of Plato and who came after Aristotle, is of particular interest. In his *Optics* and *Catoptrics* he declared himself in favour of the Pythagorean theory of the emission of rays from the eye. We should note here that there is a widespread opinion that these two books were not written by the same Euclid who wrote the *Elements*, but rather by someone else of the same name and about whom we know very little. In any case the books had a great success and an extremely long life, since seven centuries later there appeared a reprint of them edited by Theon of Alexandria who also wrote a preface to them. From this edition we are able to learn why Euclid supported the theory of the emission of rays from the eye. After remarking that someone may have a needle in front of him and not see it while searching carefully for it, he continued:

> If the act of seeing was due to the emission of images—and from all bodies images were being emitted perpetually which would excite our sense of vision—how is it that he who searches for a needle, or peruses a page of a book, does not see all of a sudden the needle or all the letters? Is it perhaps because he is not concentrating? No, since while someone searching carefully may not find a thing at once, often others, while talking, and therefore not concentrating on the task in hand, find the object more quickly. Is it because not all the images reach the eye? But why is it that not all the images which do reach the eye make an impression on it?[8]

The other argument concerns the shape of the sensitive organ. All the sense organs have a hollow form, so that they can receive; the eye, with its globular and protruding shape must be capable of emitting. We cannot honestly say that these are very powerful arguments, but this goes to show how easy it was to refute the arguments put forward in favour of the hypotheses of Aristotle, Plato and others.

Here we must discuss briefly Euclid's *Optics* since, contrary to what we find in other writers, we are faced with a text which enables us to reconstruct the author's model of light, and also because this is a very important work from the historical point of view.

It must be remembered that Euclid was a mathematician and consequently his mathematical reasoning had an important influence on the development of the theories of light. As a mathematician, and hence as a deductive thinker, Euclid required postulates in order to deduce optics. He gathered together about a dozen of them as a premise to his *Optics* and seven to his *Catoptrics*. It is not quite clear from where he took these postulates.

[8] See: Ovio: *L'Ottica di Euclide*, Hoepli, Milano, 1918

It is useful to state them here in full. For the *Optics* they are:

1. The rays emitted by the eye travel in a straight line.

2. The figure enclosed by the visual rays is a cone which has its apex at the eye and its base at the edge of the object looked at.

3. Objects on which the visual rays fall, are seen.

4. Objects which are not reached by the visual rays are not seen.

5. Objects which subtend large angles, are large.

6. Objects which subtend small angles, are small.

7. Objects which subtend equal angles appear equal.

8. Objects which are seen with the higher rays appear higher.

9. Objects which are seen with the lower rays appear lower.

10. Objects which are seen with rays directed to the right appear on the right.

11. Objects which are seen with rays directed to the left appear on the left.

12. Objects which are seen with several angles are seen more distinctly.

13. All rays have the same speed.

14. Objects can only subtend certain angles.

The postulates for the *Catoptrics* are:

1. The ray is a straight line the middle points of which touch the extremities. [This postulate is discussed later.]

2. All that is seen is seen in a rectilinear direction.

3. If a mirror is on a plane and if on this plane from any height is drawn a line at right angles to it, the line joining the observer and the mirror has the same ratio to the line joining the mirror and the height under consideration, as the height of the eye of the observer from the mirror to the height under consideration.[9]

[9] This postulate is rather difficult to understand and should be interpreted in the following manner. Let a small mirror M be placed on a plane in such a manner as to be in contact with the plane and in the same plane, and let O be a small object at a distance H_1 from the plane and M_1 from the mirror. Let the eye E of the observer be at a height H_2 from the plane and at a distance M_2 from the mirror. The postulate states that when the eye E sees the image of the object O we must have:

4. In the case of a plane mirror when the eye is on the perpendicular line joining the object to the mirror, the object cannot be seen.

5. In the case of a convex mirror, when the eye is on a line joining the object to the centre of the sphere of which the mirror is part, the object cannot be seen.

6. The same is true in the case of concave mirrors.

7. If any object is placed at the bottom of a vase, and the vase is moved away from the eye to a distance where the object is not seen any longer, the object becomes visible again at that distance when water is poured into the vase.[10]

If we ignore the great wealth of detail which is a long way from mathematical exactitude, we find in these postulates a summary of important experimental results, in addition to some hypotheses on the nature and property of light. The first postulate of *Optics* contains three important concepts, namely:

1. The concept of 'ray' as direction of propagation of light; as an elementary thread of light.

2. The concept that the light, which travels along this ray, is emitted by the eye.

3. The concept of rectilinear propagation of light.

We witness here the beginning of a definition of 'light' which even today is the foundation of geometrical optics. Had Euclid limited himself to the first and third of the above concepts, he would have laid the foundations of a theory which in its absolute clarity would have defied the centuries. Instead he wished to complete his theory with concepts of a physical and physiological nature which were unable to withstand the attack of time and of the progress of knowledge.

$$M_2/M_1 = H_2/H_1$$

[9 contd.] This geometrically means that the angle of reflection must be equal to the angle of incidence. Euclid omits to mention the fact that the two angles must also be in the same plane.

[10] See Ovio Op. Cit. p. 21 and p. 233

The second postulate is a very interesting complement to the first. It is the foundation of perspective. However, we must note that Euclid places the apex of the cone at the eye,[11] a fact which, for a geometrician of the calibre of Euclid, is rather significant. We do not know why he chose the cone when the base of the figure could have had any shape. Above all, the fact of having placed the apex at the eye means that for him the eye could be considered as a point. This remark of ours must not be considered as pedantic. It is on this and similar questions that discussions by successive thinkers took place and these discussions were of great value for the evolution of further ideas.

Postulates 3 and 4 have a physico-physiological content. The visual rays must reach an object in order that we may see it.

Postulates 5 to 11 are, on the other hand, of a physio-psychological nature, since they refer to the relation between geometrical and spatial elements (angles and directions), and to the assessment of the dimensions and of the position of the objects themselves in space.

The content of postulate 12 is purely physiological. In modern language we would say that it contains the concept of the resolving power of the eye, that is to say the acuity of vision.

Postulates 13 and 14 are only found, and then in parentheses, in the Greek text of *Optics* published by Heiberg. The first of these postulates refers to the velocity of the rays, which obviously must be understood to mean the velocity of propagation of light, and is not mentioned again in the rest of the work. The second, on the other hand, is used later in the demonstration that objects having an angular size smaller than a certain limit, are not visible.

The concept that Euclid had of light is clearly stated in the preface which, as we have already said, is attributed to Theon. It is worth examining the content of it. According to Theon, Euclid was led to the formulation of the postulate that light travelled in rectilinear rays from the observation of shadows and of beams of light through windows and slits. He saw in the rectilinear propagation of light the cause of shadows cast by bodies when they are

11 'τὴν κορυφὴν μεν ἔχοντα ἔν τψ ὄμματι'

illuminated by a source of light. Naturally he made a study of the dimensions of the shadows in relation to the dimensions of the light-source, of the opaque body and of their relative distances. There is mention even of an experiment. This consisted of two boards in each of which a thin slit was cut with a small saw and which were placed, one behind the other, in front of a small flame. The experiment showed that the light from the flame passed through the first and second slits, only when both of them were aligned with the flame. These experiments are even today mentioned in elementary text-books on optics, although experiments have shown in the last three centuries or more, that the matter is not as simple as all that.

Further on, in his preface, Theon discussed the fact that the light rays must be a little apart from each other. This rather strange idea is justified as follows:

1. An object cannot be seen at once in its whole extension.

2. Sometimes, while searching for a small object on the ground, such as a needle, we cannot see it although it is not hidden by any obstacle.

3. When we look in the direction where the object actually is, then we can see it without difficulty. Similarly, we do not see all the letters simultaneously on a written page. Theon deduced that there must be some transitory gaps in our field of vision, and explained that when we look at an object, it appears without any of these gaps because the rays explore the whole of it with a wonderful speed. After this follows a discussion of optical illusions.

Having accepted these explanations by Theon, we understand then the reasons which impelled Euclid to put forward his postulates 3 and 4 in his *Optics*, which at first might have appeared almost superfluous. On the contrary, they are necessary if we deny the continuity of the distribution of the rays in the luminous cone. Notwithstanding his rigorous logic which has made Euclid immortal, we must admit that he did not succeed in collecting in these postulates all the properties which he wished to attribute to light. In reading the text in fact, we discover some additional properties which he tacitly admitted. As an example, in the 23rd Proposition of his *Catoptrics* he spoke of longer and shorter rays and of the subsequent perception of further or nearer points, as if the eye could feel the length of the rays which it emitted.

The most remarkable example of this is the difficulty of reconciling the postulates of his *Catoptrics* with those of his *Optics*. When he discussed reflection, Euclid seemed to forget readily that the first postulate of *Optics* demanded that the rays emitted by the eye had to travel in a straight line and he offered no explanation why we should make an exception when such rays hit a mirror. Why did he not make this reservation when he stated his first postulate of *Optics*? Moreover, the first postulate of *Catoptrics*, which states that 'the ray is a straight line the middle points of which touch the extremities' must be interpreted in the sense that the straight line may also be a broken line, otherwise the postulate cannot be reconciled at all with the phenomenon of reflection to which in particular this postulate refers. Where Euclid revealed a touch of genius is in his second postulate of *Catoptrics*: 'all that is seen is seen in a rectilinear direction'. We could venture to say here that he had an intuition of the difference between the light ray and the visual ray, and that he had realized the difference in their behaviour, the light ray being bent on the mirror, while the visual ray never bends. He expressed this second concept very clearly in his second postulate, while he appeared to have had neither the capacity nor the courage to express the first. Since he had accepted the emission of rays from the eye, this division of the ray into a light ray and a visual ray, must have appeared to him very difficult to admit; hence the rays must have been united until they met a mirror and then one was deviated and the other was not! Faced with this difficulty Euclid preferred to keep silent; an attitude which is not the prerogative of ancient science alone.

The third postulate of *Catoptrics* contains the law of reflection on a plane surface, a law later also applied to both concave and convex spherical surfaces. Postulates 4, 5 and 6 are rather strange. They repeat the same concept for plane, concave and convex mirrors, that is to say that when the eye is placed on a perpendicular line joining the object to the mirror, the object cannot be seen. The seventh postulate is an experiment of refraction.

The two extremely interesting books which Euclid wrote on

these subjects are fundamentally books of Geometry. Therefore *Optics* has also been called *Perspectiva* and its content is of great value, but if we are to understand it we must eliminate a whole mass of concepts, of unrelated and often incoherent ideas which are physical, and physiological as well as psychological. There is no doubt that this book is the result of an experimental study which was brilliant rather than rigorous. The newly emerging science of optics is presented in all its complexity. Man, in his search for knowledge fashioned for himself a powerful weapon in order to begin the work of discovery. He created the theory of a 'ray'. This is a mathematical abstraction which was necessary if the human mind was to advance in this field. With this simple ideal 'light', obedient to impossible laws, man created the means by which he could reason and progress experimentally as well as theoretically.

This theory, this creation, now superseded but still alive after over a thousand years, is Euclid's most valuable contribution. He created this theory, and although he made it function admirably both in his *Optics* as an emitting element, and in his *Catoptrics* for the study of reflection, it remained surrounded by additional ideas with the appearance of physical characteristics. Although Euclid was a genius, we cannot say that he reached noteworthy conclusions. We are faced, in reality, with a mass of disconnected and inconclusive observations.

Nevertheless, some of his points deserve our attention. Euclid maintained that light was emitted by the eye, and yet, if we accept Theon's preface, he postulated that light travelled in a straight line on the basis of observations and experiments carried out with external sources of light. It would be very interesting to know how he reconciled the two types of light. Did he believe that the light emitted by the eye and which reached the objects was of the same nature as that emitted by the sun which entered a room through a window? This is not easy to answer. The last Proposition of *Catoptrics* however, the thirty-first, says: 'with concave mirrors directed to the sun it is possible to light a fire'. In the proof that follows he discussed the rays emitted by the sun which are reflected by the mirror, in the same way as in the previous proof he discussed the rays emitted by the eye which are reflected by surfaces until they reach various objects. On the other hand this concept was so deeply rooted in his mind that it led Euclid to accept as true some conclusions which are almost absurd, for example we may quote part of Proposition 24: 'If the eye is at the centre of a concave mirror, it can only see itself'. It could not have been otherwise if the rays which left the eye reached the mirror, since all of them, after being reflected, had to fall back on the eye itself. This conclusion could have been proved absurd experimentally with the means which Euclid appeared to have at his disposal. Furthermore, his own writings contain basic ideas which should have been sufficient to prove to him its absurdity. In fact in the third Proposition of his *Optics* we find:

> It is necessary that between the eye and the object there should always be a certain distance, otherwise the object cannot be seen.

We see here the embryo of the idea of near point or 'puntum proximum'. Moreover, the rigorous application of the second postulate of *Catoptrics* should have led to more accurate conclusions.

The complexity of the subject fully justifies these deficiencies. The human mind, even one as great as that of Euclid, could not have achieved the results which were to be obtained by the work of many scientists during many centuries.

Euclid has the merit of having created the rectilinear ray, which had no physical substance and which has served so well as the basis of Geometrical Optics up to the present day. In addition we find in his work the foundation of perspective, the laws of reflection and of the images both for plane and spherical mirrors. We also find many observations of physical, physiological and psychological optics, though unrelated and sometime even misinterpreted. Nevertheless, all this is sufficient to place Euclid among the greatest contributors to the science of optics.

There is no doubt that Euclid's contributions played an important part in the development of optics. A new edition of his work was published after seven hundred years and this was done not for historical reasons but because of its value as a text-book. Among his followers we find famous men such as Hipparchus of Nicaea (*c.* 190–120 B.C.) who, if we can trust Plutarch, believed that the rays emitted from the eyes of a person went out to touch the objects like hands at the end of arms, and brought back the elements necessary for seeing the objects.

The famous Ptolemy of Alexandria (*fl.* A.D. 170), who substituted a pyramid for Euclid's cone of perspective with the apex at the eye, thought that the rays of light were emitted by the eye which felt their direction and length. He believed that the middle rays of the pyramid of perspective had a greater visual capacity. He also studied binocular vision and the doubling of images when the pyramid of the right eye is not correctly combined with that of the left eye. He considered that colours were a superficial layer on the bodies, and he studied the question of evaluation of the size of observed objects by combining the height of the pyramid of perspective with the dimension of its base. Ptolemy too, has left us a treaty on *Optics*, in which, among other things, there is a noteworthy and accurate study of the images produced by both concave and convex mirrors. This study must be considered much more advanced than that which is described by Euclid in *Catoptrics*. In addition, the results of his measurements, made in order to establish the law of the phenomenon of refraction, are particularly interesting. He formulated a law of refraction but in a form which later proved to be inaccurate. Although in his work there is little of value concerning the nature of light, what we do find of great interest is the improvement of experimental investigations, and as a result the increase in the knowledge of the properties of light. On the whole, however, we still find a chaotic mixture of geometrical and physical ideas, as well as physiological and psychological ones, and there is no synthesis of any value.

In a third ancient text-book on optics, written by Damianus or Heliodorus of Larissa, whose date is uncertain but later than Ptolemy, we find a very important contribution.

This author too insisted that rays were rectilinear and that the structure of perspective vision was conical. He realized, however, that not only the eye but even the pupil was too large to act as an apex of the cone, and without explaining why, he assumed the apex of the cone to be within the eye, at the centre of a sphere of which the pupil represented a quarter. This first step within the membranes of the globe of the eye will have important consequences as we shall see later.

Another point of interest is the comparison that Heliodorus made between the rays emitted by the eye and those emitted by the sun, a subject which, as we have already mentioned, neither Euclid nor any other optician of his school had considered. Heliodorus showed that both these rays had the same experimental properties.

Heliodorus made a deeper analysis of the hypothesis of the emission of rays. As we have already mentioned, he assumed that rays were emitted by the eye, but added that the true hypothesis consisted in the admission that the rays were rectilinear and in the acceptance of the law of reflection, since both enabled him to make a study of the images. In addition, he stated that vision occurred directly by reflection and refraction, that the rectilinear path of light was a necessary consequence of the fact that the path had to be the shortest possible:

> Because if sight must reach the object to be seen in the shortest possible time, then it must travel along a straight line, since this is the shortest line joining two points.[12]

He continued by reporting the conclusions reached by Hero of Alexandria (*c.* A.D. 100) concerning the shortest path followed by rays also in the case of refraction. Hero proved in his book on mirrors, that lines which on meeting a mirror formed equal angles, were shorter than all other lines which originated from the same points, but which when meeting the mirror formed unequal angles. Having proved this he stated:

> If nature is not to operate in vain with the rays of our sight, then the bending of the rays must occur in equal angles. This is clearly seen in the case of the rays of the sun which are bent at equal angles.

[12] See *La prospettiva* by Heliodorus of Larissa, translated by P. Egnatio Danti, Florence Giunti, 1573

Had Hero fully extended to refraction this law of the shortest path he would have been known as the discoverer of what is today known as Fermat's Principle.

Although several historians believe that Heliodorus has taken all his ideas from Hero of Alexandria, who probably lived in the first or second century A.D., we must admit that there is a remarkable maturity of thought in all these observations. It is because of this that Euclid's Geometrical Optics was recognized, but remained mixed and confused with other optics which were still indefinite and unformed.

The stage we have reached in the development of ideas on the nature of light can be described briefly as follows. A considerable number of various types of fundamental ideas, namely mathematical (geometrical), physical, physiological and psychological, had been collected together. The geometrical elements, which were more abstract but also more mature, acquired a particular importance which was much superior to all the others. Leaving aside the hypotheses and considerations of a physical, physiological and psychological nature, light was studied only in its geometrical elements, which were represented by the rectilinear structure of the visual rays, by the law of reflection, by the bending of the rays by refraction and by all the phenomena of perspective. This optics can stand alone and can progress from these fundamental hypotheses. It is geometrical optics, which is still very much alive and in use today, nearly two thousand years after Hero and Heliodorus had cleared away the useless principles which surrounded it.

We have now reached the third century A.D. and the theories connected with the emission of rays from the eye were still very much alive. Only then, after nearly seven centuries, was there a realization that of all the conclusions reached so far, only geometrical optics was really important and that it was true whether the light was emitted by the eye or came from an external source. Neither of these two hypotheses received confirmation or rejection from the studies of geometrical optics. This is surprising if we bear in mind that at the same time there had developed studies following the lines suggested by the atomistic school according to the direction of ideas attributed to Leucippus.

A very similar concept, but better defined, is found in a letter from Epicurus of Samos (342–270 B.C.) to Herodotus, and which is reproduced in Book X of Diogenes Laertius' work. We give here the interpretation of a very controversial passage as given by Trouessart[13] according to the Schneider and the Hubner editions:

> From all existing bodies are emitted continuously certain forms which have an appearance similar to these solid bodies, but are much more tenuous than anything that can be felt. . . . These forms are what we call *eidola* (ειδολα) or images. Their motion, which takes place in a vacuum, without any resistance, has such a speed, that it enables them to cover any imaginable distance in a negligible time. . . . We must also remember that these images are formed at the same time as the thought is born. Because of the continuous emission by the surfaces of bodies, there are emitted also images which are not perceived. These emissions maintain the relative position and order of atoms for a long time, although occasionally some confusion may occur.
>
> On the other hand these are formed without delay in the air, because they are forms without bodies. . . .
>
> In all this there is nothing contradictory to the functioning of the senses, if we take into account the manner in which these senses must be stimulated in order to give us impressions which are in harmony with the external objects. We have to admit that we see the forms of the external objects and that we think as a consequence of 'something' of these objects which has entered into us. External things cannot reveal their proper nature, for example their colour, by means of the air which is interposed between them and us, nor by means of rays nor by any other similar emission which would travel from us to these objects, in the same way as this can be done by images sent to us by these objects; images which, while they maintain the similarity of the objects in colour and form, enter into our sight and into our thought according to the proportion of their size and with a very rapid motion. The image φανταισα (Fantaisa) of the form and of the characteristics which in this way we receive by impression on our spirit or on our senses is actually the form of the solid (of the real object). . . .

[13] Op. cit. p. 170

The deception and the error is caused by 'opinion' which is added, following an action occurring within ourselves and which unites with the notion supplied to us by sight. We must, therefore, restrain our opinion and prevent its intervention from spoiling everything.

The same happens in the case of hearing, which takes place when a body emits a word, a sound, a whistle or something capable of giving us the impression of sound. . . . We must say the same thing for the sense of smell, that is to say that no impression can be made on this sense except by means of certain corpuscles which are given off by the thing which is smelt and which have a certain size of suitable proportions to stimulate this sense. Thus among these corpuscles, some produce their impression in a confused and unpleasant manner while others in a more orderly and pleasant manner.

Although it had a lesser effect than the Pythagorean theories, this idea of the emission of simulacra, of shadows, of images by the bodies lasted a long time since it was strongly supported by the Epicurean school. A very important text on this question is *De Rerum Natura* by Lucretius[14] who was a staunch supporter of the Epicurean doctrines. In the first part of Book IV of his poem many optical questions are discussed and in particular that of vision. This is how he expressed himself:

I will begin to tell you what exceeding nearly concerns this theme, that there are what we call idols of things, which may be named, as it were, films or even rind, because the image bears an appearance and form like to that, whatever it be, from whose body it appears to be shed, ere it wanders abroad.

First of all, since among things clear to see many things give off bodies, in part scattered loosely abroad, even as wood gives off smoke and

[14] Where Lucretius was born is not known with certainty but probably in Rome where he lived and died. The year of his birth is placed between 99 and 95 B.C. and that of his death between 55 and 51 B.C. *De Rerum Natura* is his only work of which we know something and even this work remained unfinished on account of the death of the author. The book has had many editions and translations particularly during the Renaissance.

In our book we shall refer to the English translation by Cyril Bailey—Oxford University Press, 1947

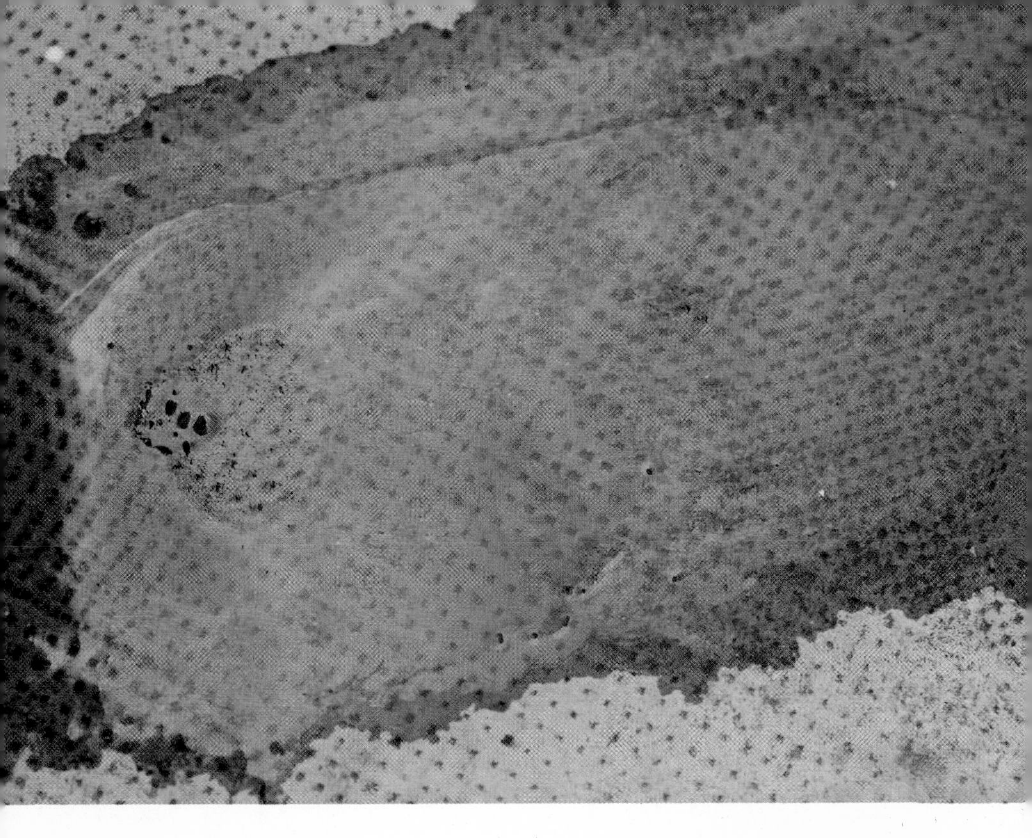

A. Colours produced by a thin film of oil on a wet surface. This effect was first described by Hooke in *Micrographia* in 1665. Today, the thickness of the film of oil is known to be of the order of 0.001 millimetre.

B. Solar spectrum as observed by Newton and as described by him in *Opticks*.

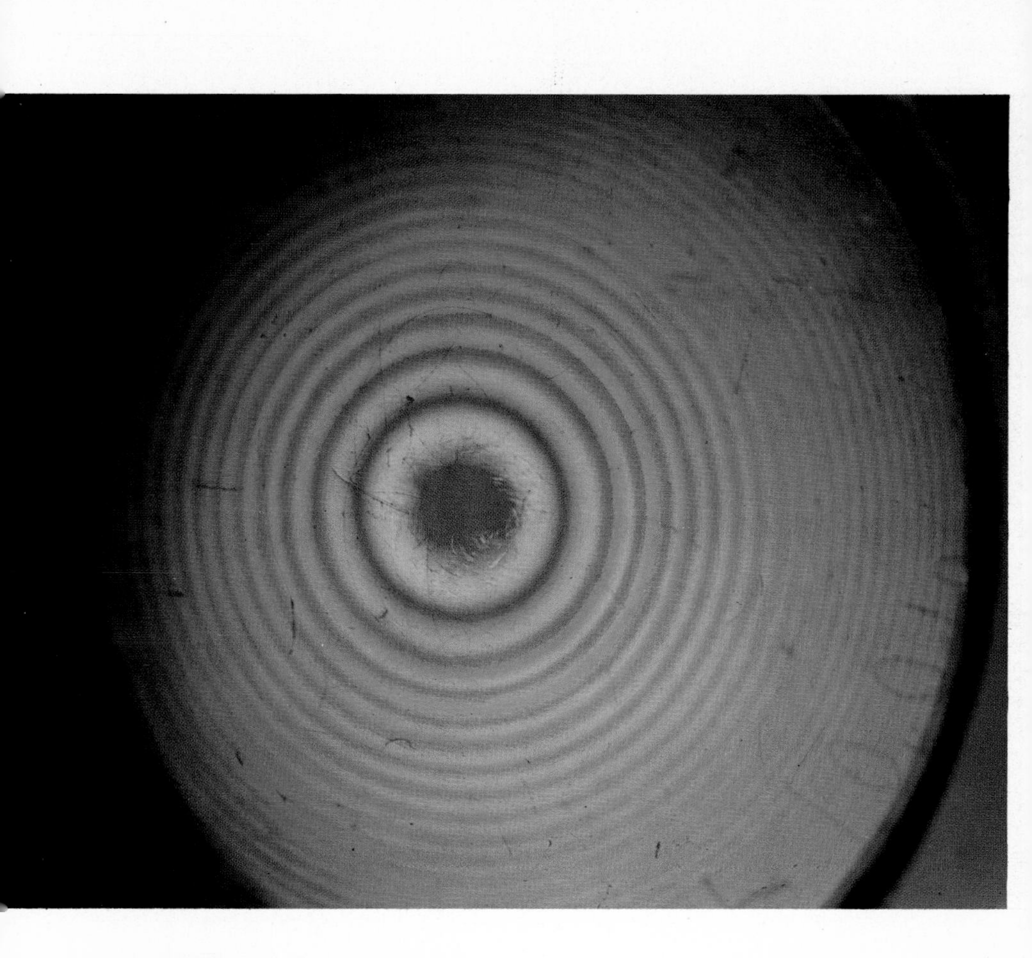

C. Rings formed between a plane surface and a spherical convex surface of very small curvature. Described by Hooke in *Micrographia* and later studied by Newton who carried out accurate measurements. Known today as 'Newton's Rings'.

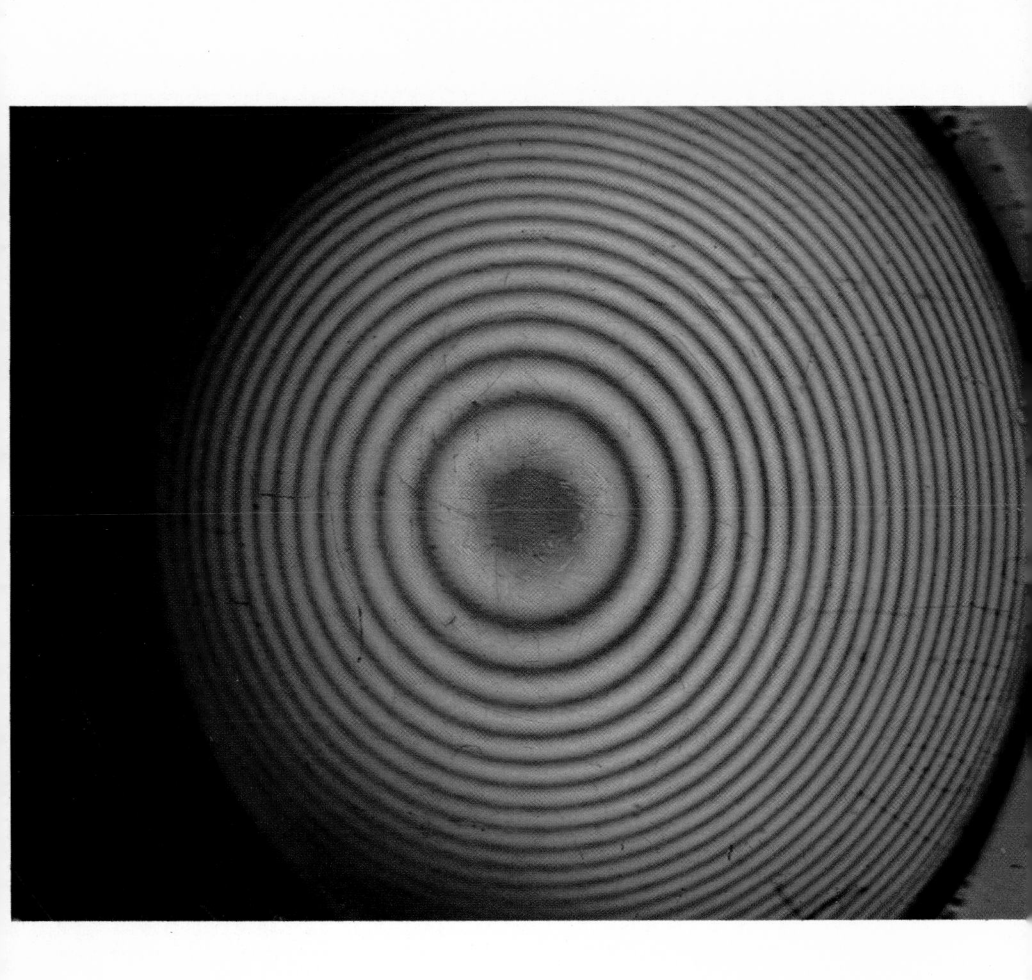

D. Rings similar to those of Plate C but produced by monochromatic light of sodium. Note how many more rings are visible when this light is used.

E. Interference fringes caused by the overlapping of the two 'Airy's discs' (see frontispiece) that are produced by two pinholes very close to each other and less than one millimetre in diameter. This was one of the most interesting of Young's experiments.

F. Interference fringes obtained by means of 'Fresnel's Mirrors'.

G. Interference fringes obtained by means of 'Fresnel's Biprism'. The colours appear different when compared with those of Plate F. This is due to the dispersion of the prism.

fires heat, and in part more closely knit and packed together, as when now and then grasshoppers lay aside their smooth coats in summer, and when calves at their birth give off cauls from their outermost body, and likewise when the slippery serpent rubs off its vesture on the thorns; for often we see the brambles laden with wind-blown spoils from snakes. And since these things come to pass, a thin image from things too must needs be given off from the outermost body of things. For why these films should fall and part from things any more than films that are thin, none can breathe a word to prove; above all, since on the surface of things there are many tiny bodies, which could be cast off in the same order wherein they stood, and could preserve the outline of their shape, yea and be cast the more quickly, inasmuch as they can be less entangled, in that they are few, and placed in the forefront.[15]

Lucretius then quoted an obvious case. In a theatre the bright coloured awnings tinge with their colours the people and things which are underneath. If then this colour is given out from the canvas in order to affect the neighbouring objects, why should not this phenomenon be accepted generally? And Lucretius concluded:

There are then sure traces of forms, which fly about everywhere, endowed with slender bulk, nor can they be seen apart, one by one.[16]

He then compared light with smell, heat, smoke and other emanations of the same type, and he concluded that while the first had no obstacles on account of its extreme delicacy, the others were diffused and dispersed, because they could not find 'straight outlets to their paths'.

Finally he found that since simulacra seen in water, in mirrors and in all shining bodies were perfectly similar to the real objects, they could only be formed by their images, which as described above, were emitted by the same bodies. All bodies then emitted similar images which could not be seen separately, but which when reflected and re-united by means of mirrors could reach our eyes.

[15] Op. Cit. Book IV, vv. 46–70, page 364

[16] Op. Cit. Book IV, vv. 87–89, page 366

Later, Lucretius considered the great velocity with which these images travelled through space. The arguments and experiments he discussed on this point appear to us very naive, but in reality their only purpose was to give us some understanding of the question. Finally a kind of peroration completed the general picture.

> Wherefore more and more you must needs confess that with wondrous swiftness there are sent off from things the bodies which strike the eyes and awake our vision. And from certain things scents stream off unceasingly; just as cold streams off from rivers, heat from the sun, spray from the waves of the sea, which gnaws away walls all around the shores. Nor do diverse voices cease to fly abroad through the air. Again, often moisture of a salt savour comes into our mouth, when we walk by the sea, and on the other hand, when we watch wormwood being diluted and mixed, a bitter taste touches it. So surely from all things each several thing is carried off in a stream, and is sent abroad to every quarter on all sides, nor is there any delay or respite granted in this flux, since we have sensation unceasingly, and we are suffered always to descry and smell all things, and to hear them sound.[17]

If we wish to know the reasons why the supporters of the theory of simulacra appeared to be convinced of the existence of these strange entities, we ought carefully to analyse Lucretius' book which, although very concise, gives us important information on this particular question. For example, it is extremely interesting to find the idea of the formation of simulacra linked to an atomic structure of matter. According to Lucretius everything was made of extremely small units which he calls *primordia* and *exordia rerum*. The amorphous exhalations of substances such as those affecting the sense of smell, for example, were composed of these *primordia*. Lucretius pointed out that these exhalations were formless because they emerged from the interior of bodies, and hence, in order to be able to pass through the material pores, they lost their natural order. On the other hand when they detached themselves from the surface of a body, they were not disturbed by having to pass through anything and therefore they could maintain the

[17] Op. Cit. Book IV, vv. 216–29, page 372 and 374

order they had on the surface from which they detached themselves. This is how simulacra were formed. Lucretius however, did not mention how they could penetrate through the human pupil without having to contract.

On the other hand there is a remarkable idea which, as far as I know, is expressed only by Lucretius. One of the most difficult problems to solve was how the eye could determine the distance of the emitting body using the information imparted to the eye by the simulacra which had entered it. This information was absolutely necessary for the soul (*psyche*) to reconstruct a representation of the original body and to enable it to acquire a definite knowledge of the proper dimensions and place of the particular body in the apparent world. Lucretius was courageous enough to face this problem and he said:

> And how far each thing is away from us, the image causes us to see and enables us to distinguish. For when it is given off, straightway it pushes and drives before it all the air that has its place between it and the eyes, and thus it all glides through our eye-balls, and, as it were, brushes through the pupils, and so passes on. Therefore it comes to pass that we see how far away each thing is. And the more air is driven on in front, and the longer the breeze which brushes through our eyes, the farther each thing is seen to be removed. But you must know that these things are brought to pass by means exceeding quick, so that we see what it is and at the same time how far it is away.[18]

Lucretius was so convinced of this mechanism, that he applied it also to the case of reflection by mirrors. He explained the fact that we could see an image behind a mirror at the same distance as the object was from the front of the mirror by means of the *breeze* which he mentioned in the quotation above. The *breeze* was also reflected by the mirror and therefore determined the distance of the image from the eye. Evidently this explanation was not accepted by the supporters of the theory of simulacra and has been completely forgotten. It was, however, an important proof of the efforts made by the philosophers to explain the details of optical phenomena.

[18] Op. Cit. Book IV, vv. 244–55, page 375

Lucretius' reference to the great velocity of simulacra is much more interesting. Let us, however, return to the question of the history of light and in this connection we must note that the terminology used by Lucretius is very varied and rather indefinite. For simulacra we find the words *effigiae*, *figurae*, *imagines*, *membranae*, and *cortex*. It is true that Lucretius was writing in verse, but this variety of terms cannot always be justified by the exigence of the metre. Rather this philological variety must be accepted as a proof of the fact that the ideas were not very clearly defined.

Another very important point is the implicit hint that Lucretius gave of the necessity of an agent to produce the emission of simulacra by bodies, since obviously these bodies did not emit anything which revealed them when in the dark. Lucretius this time did not face the problem squarely, but in his writings we catch a glimpse of his intention to consider that such an agent should be a 'something' (*quid*) which was emitted by the sun. Here again the variety of the terms used indicates that the ideas are not really clearly defined. Lucretius in fact made use of *lux*, *lumen*, *fulgor*, *splendor* and *clarus candor*. We can say that generally Lucretius used the word *lux* to indicate the condition of the surroundings, although sometimes he used the expression *solis lux et vapor* with the same meaning. *Fulgor* and *splendor* he used for brightness. Much more precise however, is his usage of the word *lumen* as something emitted by the sun to fill space. So he said:

> Just as the sun has to emit many *lumina* in a brief space of time, so that space as a whole may be constantly filled. . . .[19]

Lucretius tried to justify the great velocity of propagation of simulacra and he said:

[19] *De Rerum Natura*, Book IV, vv. 161–2.
Note that in our translation the word *lumina* has been kept to indicate that it is extremely difficult to know what meaning to give to such a word. Any modern word would probably give an interpretation which may not in effect correspond to the intentions of the author.

> *Lux* and *vapor* are of this type because they consist of very small particles which are pressed forward and stream unceasingly through the interposed air, impelled by the pressure of the particles that follow. For immediately *lumen* is followed by other *lumine* and *fulgur* is spurred on by *fulgere*.[20]

Later when Lucretius described the shadows produced by the sun on the ground, he took them to be regions devoid of *lumine*.[21]

Although the idea is not expressed explicitly, there is no doubt that in these pages of Lucretius there is already more than a hint that *lumen* was something emitted by the sun and consisting of extremely minute particles which were launched into space and which rapidly filled it completely. It is an embryo of an idea, which as we shall see later will give rise to very important developments. Following the same line of thought Lucretius gave us an important contribution. He said:

> Bright things moreover the eyes avoid and shun to look upon. The sun too blinds, if you try to raise your eyes to meet him, because his own power is great, and the idols from him are borne from on high through the clear air heavily, and strike upon the eyes, disordering their texture. Moreover, any piercing brightness often burns the eyes for the reason that it contains many seeds of fire, which give birth to pain in the eyes, finding their way in.[22]

This is very important, because as we shall see, the idea expressed was taken up again and given more prominence a thousand years later. We may note that Lucretius here discussed the *idols* of the sun, which were so intense as to burn the eyes, while somewhere else he discussed the *lumen* emitted by the sun, and he was very careful not to identify the two concepts. It is very interesting to note that while observations of this type were made and even incorporated in a poem, the Pythagorean philosophers ignored

[20] *De Rerum Natura* Book IV, vv. 185–90

[21] Op. Cit. Book IV, vv. 364–74

[22] *De Rerum Natura*, translated by C. Bailey p. 379, vv. 324–31

them completely and continued to think and to talk in terms of rays which, since they originated in the eye, could certainly not be the cause of the burning of the eye itself. This line of thought persisted for over a thousand years. This is a proof of the great prestige of abstract concepts in which mathematics is so rich.

We have quoted these passages from Lucretius to show that there existed an extremist current of thought which admitted no compromise. According to it, vision took place by means of the emissions from bodies of images identical to the bodies themselves and directed to the eye. These extremists however, did not have many followers. Even in their own time they must have appeared as fanatics because their reasoning was too narrow and was open to criticism. Moreover, not all the followers of these theories were as fixed in their ideas as Lucretius, even Epicurus, for example, had some very strong reservations. He admitted that 'opinion' might be mistaken and had to be 'controlled' in order that its intervention should not affect the results of vision. Hence images originated in the bodies, they carried the forms and the colours, but then vision could occur in a quite different manner!

In order to show the perplexity which existed at the beginning of the Christian Era, we shall report here the criticism which is to be found in Macrobius.[23] In the dialogue between Disarius and Eustaphius the former quotes the theory of the emission of images which he attributes to Democritus and Epicurus. Here is the relevant part of the dialogue:

> DISARIUS: . . . Epicurus believes that certain images continuously detach themselves from all bodies. Although it is impossible to attribute to their trajectories the shortest interval of time, these images, which are empty figures, carry the coherent husks of the bodies, are received by our eyes and so reach the seat of the sense which nature has assigned to them. This is what this philosopher maintains, if you have different views I am ready to hear them.
>
> EUSTAPHIUS (laughing): It is easy to see what misled Epicurus. By following an analogy with the other senses, he has strayed far from the truth. In fact when we hear, taste, touch or smell an odour we do not

[23] *Saturnalia*, Book IV, Chapter XIV

emit anything but on the contrary we receive from the exterior 'something' which stimulates the corresponding sense. . . . This is the reason that led him to think that nothing emerged from the eye but rather that the images themselves penetrated our eyes. This opinion, however, is contradicted by the fact that he who looks at his own image in a mirror, sees it as if the image had turned in order to look at him. Since the image has parted from our body in a straight line, we ought to see the back of it, so that the left should correspond to the left, and the right to the right. In fact the actor who removes his mask sees it from the side which had covered his face, namely he sees not the outside but the inside of the mask.

I would also like to ask that philosopher whether the images detach themselves from the bodies only when we look at them, or whether the images pour forth on all sides, even when no one perceives them. Because if he maintains the first opinion, then I would like to ask him what is the power which rules these images, so that they are always available to him who looks and when he turns his gaze away from them, so do the images turn. If on the other hand he prefers the second proposition, namely that images detach themselves continuously, I would like to ask him for how long would their parts remain coherent, since they have no ties to hold them together, and even if we agreed that the images could hold together, how could they maintain a certain colour whose nature although disembodied, cannot exist without the body? Moreover, who can believe that as soon as he turns his eyes to one side, all the images will immediately come towards him, images of the sky, of the earth, of the rivers, of the plains, of the ships and of many myriads of things which we see at a glance, particularly since the pupil which allows us to see is extremely small? And how can we see a whole army? Will the images which detach themselves from each soldier group themselves and enter the eye of the observer aligned as the soldiers are? *But why waste words in fighting an opinion which is so inconsistent that it refutes itself by its own vacuity*?

It is easy to demolish, but when it was a question of reconstruction, even Eustaphius would probably no longer have smiled. Indeed, he must have found it extremely difficult to collect enough ideas to launch a theory of a Platonic type and to conclude:

Three things therefore are necessary in order that we should be able to see: the light which we emit from our interior; the air interposed and which must be luminous and the body which for its part terminates the *tension*.

As can be seen, here we find a considerable confusion of ideas.

In conclusion, at the end of this first period of the study of the phenomenon of vision and of light we find that some facts and some results deserve special mention.

1. There is a considerable number of experiments in optics in the widest sense of the word, mainly in the field of vision but also outside this field.

2. We have found the development of three theoretical lines of thought. First that of the Pythagoreans who considered that rays emerged from the eye. Secondly, that of the Atomists who suggested that bodies emitted simulacra. Thirdly, that of the Platonic school which attempted to combine both the emissions. A fourth theory, which does not appear to have much following although it was supported by Aristotle, considered vision as being the result of a movement between the object and the eye.

3. The theory of the Pythagoreans seems to predominate and to have greater vitality than the others.

4. This theory becomes part of Geometry, with the consequent definition of 'visual ray' which is emitted by the eye, which follows a straight line and which can be reflected and refracted.

5. Geometrical Optics and Perspective were born, together with the laws of reflection and with important premises of the law of refraction and of that of the minimum optical path.

6. The idea, even if only in embryo, appeared that from luminous sources a 'something' was emitted to which was given the name of *lumen* and which consisted of myriads of extremely minute and fast particles, capable of filling space completely and capable also of acting upon the eyes to such an extent as to hurt them.

This is indeed a remarkable collection of ideas. The character of this first period of investigation is that usually found when research is confronted by a very complex subject. It consists of a multiplicity of notions, of terms, of observations and of conclusions and in consequence there is a certain confusion. Nevertheless, such a period is a vital one because, from the study of this early experimental and rational material, the future investigators will be able to bring some order where confusion existed; they will be able to introduce classification of facts, selection of important experiments and formulation of theories. The optical phenomenon of vision is fundamental and extremely complex and it was logical that the study of optics should begin with it. Geometrical, physical, physiological and psychological factors enter into the phenomenon of vision and to these there correspond as many branches of optics, each one vast and rich in theories, experiments and indeed in mystery even to this day.

Thus in this first period of the history of light, we witness an indistinct mixture of these four branches of optics. We also witness the separation of the first of them, namely of Geometrical Optics, from all the others. Generally speaking, in the rest there remains a mass of concepts and of studies of a psychological nature. The first object of study, as is almost always the case, was the individual himself; namely, the mathematical thoughts and the psychology of the individual who was carrying out the investigations. Many more centuries were needed for the investigations to be extended to the world surrounding the individual and to the greater external world, and, as a result of this, for the ideas concerning light to become clearer and more defined.

CHAPTER TWO

Optics in the Middle Ages

IF we are to understand the evolution of ideas concerning the mechanism of vision and the nature of light in the centuries which were to follow, we must consider some of the weaker points of the theories discussed in the previous chapter. It is a question of ideas implicitly accepted and understood without discussion. Had these ideas been analysed carefully and critically, there is no doubt that the weak points would have been detected, particularly since the philosophers interested in this subject showed great wisdom and perspicacity.

The development of ideas was for a long time delayed and wrongly directed by the first assumption that the seeing of a body was to be considered as a single, indivisible operation. An object either large or small had to be seen in its entirety and in the texts that have reached us there is no indication of the thought that seeing an object might be the cumulative effect of the operations necessary to see each of its elementary parts. It seems that all philosophers thought that the larger an object was the easier it was to see it. All of them appear to have avoided examining the case of seeing a star or a minute object as if it were a limiting case and only to be considered after having solved the main case worthy of interest, namely that of a solid body subtending a large angle.

The theme, which is repeated explicitly and very decisively by all, is that vision takes place by a cone or a pyramid which has its apex in the eye and its base on the object observed. This is the concept of perspective *par excellence*. Evidently the efforts of the theoreticians were concentrated on letting the 'husks' or images enter the eye and meanwhile no one worried about the limiting case in which the cone might have a very negligible angle at the apex. The tendency was to widen the base of the perspective cone and to reduce the apex until the whole eye became a point. Mathematically this was a very useful simplification for the theory of perspective, but at the same time it had dangerous consequences for the theory of vision as a whole, for difficulties are not solved by ignoring them. This was in fact a second error in the approach to the problem. Instead of investigating the structure of the eye in order to explain the causes, namely to deduce the nature of the stimulus as Empedocles had already attempted to do, philosophers ended by seeing the eye only as a sensitive point-like organ whose dimension and structure were negligible for the function which it had to fulfil.

It is not possible to follow the development of ideas for centuries after Ptolemy because of lack of detailed texts. In earlier centuries, and indeed in more modern times, texts followed each other but in the period under discussion we find only now and then and at very great intervals of time, some text which by its explicit content and by its language shows that in the meantime both experiments and theories had progressed, but it is extremely difficult to know to whom merit is due.

A decisive contribution to the subject was made by Galen (A.D. 131–201) with his physiological and anatomical discoveries. Of special interest is his description of the structure of the eye which he included among all the sensorial organs, emphasizing the function of the optic nerve in the case of vision. Galen went as far as seeing in the nerve the most important element of vision itself. He thought of this nerve as a channel through which passed a visual fluid originating in the brain, by analogy with similar fluids which were thought to reach all the organs of the senses in order to render them sensitive. The visual fluid would be diffused in the vitreous humour through the retina, which forms the back of the eye, and then would reach the crystalline lens and render it sensitive, that is to say responsive to the light arriving from the exterior. As for light itself, Galen did not attempt any explanation. He considered it to be a fluid diffused by the sun into the air, just as the brain injected its fluid into the nerves so that in effect, air performed the same duty for external light as the nerves did for the visual fluid. Without trying to define the nature of the external fluid, Galen limited himself to considering its rectilinear propagation which by now had been consolidated by geometrical optics, and he discussed the perspective cone according to which vision took place. Finally, we can say that Galen returned to Platonic ideas, that there was an external fluid which travelled towards the eye and an internal fluid which, although not emanating from the eye, made it sensitive and likely to respond to the external fluid. Here we begin to find the first indication of the part played by the structure of the eye itself even if in this case the concept, that the crystalline lens is the sensitive part, is false.

These views had a very decisive influence upon the development of the theories about the mechanism of vision and hence of light. The unsettled historical period which followed the times of Galen resulted in a move to the Middle East of the cultural centres which were interested in this type of speculation. While the western philosophical and scientific world was undergoing a very serious crisis, there developed in the Islamic world centres of cultural activities which were worthy of note. It is very unfortunate that the texts relating to these activities present serious difficulties, first because only a few still exist and secondly because most are still today only found written in Arabic. The only information that we have available and that we will discuss is mostly second hand. It was either obtained from Latin translations of the original Arabic texts that were made much later, or from comparatively recent translations by language experts. These experts are no doubt very eminent and very reliable but, as extremely competent interpreters of the original language, they approached the work with a philological and humanistic background and found considerable difficulty in faithfully interpreting the thoughts of the authors who not only wrote in the Arabic of ten centuries ago, but also used what we could today call technical terms, the accurate translation of which requires a profound knowledge of the subject. What we are to discuss here as the contribution of the Schools of the Middle East to the question of light is bound to be somewhat brief. These few ideas however, will be sufficient to show how much attention was given to the questions connected with the main subject of our book, in the centuries around A.D. 1000.

We ought to remember first of all, that in this period and especially in the ninth century A.D., there existed in Baghdad a flourishing and extremely important medical school. This school was particularly interested in the structure of the globe of the eye and of the visual apparatus in general. All this led to important discussions as to the seat of visual sensations. In this field the works of three Arab physicians were extremely influential during the Middle Ages.

Hunain Abn Is-haq (809–873), who became well known for

translating Galen's works from Greek into Syriac, became known in the western world under the latinized name of Johannitius. At the same time Hubaish, a collaborator of Hunain, was translating Galen's works from the Suriac into Arabic.

In the following century we find two more important writers; Ali Ibn Isa, also known as Jesus Hali and who was considered as a successor to Hunain, and Rasis (Abu Bekr Muhammed Ibn Zakariya Al Razi), who was a famous leader of a school in the tenth century. The works of these leading teachers of medicine were translated into Latin. One of the most interesting is *De Oculis* which for a long time had been attributed to Galen but which in fact contains the greater part of Hunain's ideas. Another very important Latin translation was that of a really imposing work, a medical encyclopaedia in 25 Books written by Rasis and which became known as *Continens Medicinae*. During the Middle Ages these works assumed the character of classic texts. Their main concern was medicine but in them is found an extensive study of the anatomy of the eye and some attempt to advance a theory of vision. These theories however, are very similar to those advanced by Galen and are not strictly relevant to the history of light. Nevertheless the interest centred on the eye, on its anatomy and on its physiology, was bound to lead to a consideration of the phenomenon of vision in relation to physico-physiological rather than geometrical factors.

The necessity of giving to the various theories concerning vision, a less mathematical and more physico-physiological direction was asserted in a very explicit and detailed manner by another famous Arab scientist, Abu Ysuf Yaqub Ibn Is-haq, better known under his abbreviated name of Alkindi. He lived in Basra and Baghdad between A.D. 813 and 873. He became very famous and was known as the first Arab philosopher. His scientific work ranged over a very wide field and his influence spread over many branches of physics. The research on his work carried out by students of the history of science is in itself extensive and extremely important.

As far as we are concerned, out of all his works two are of interest to us, those which in the Latin translation have the titles of *De*

Aspectibus and of *De Radiis Stellatis*. The latter, which is still in manuscript, is a work with an astrological background, as it deals with the stellar influences upon the terrestrial world. What interests us, however, is that in it we find a confirmation of the concept that it is not only the stars which act in all directions by means of *radii*, which are to be understood as radii in the geometrical sense of the word, but also all the things of this world. Material bodies (*substantiae*) as well as immaterial bodies (*accidentes*) including the voice, all act exactly in the same way as the stars.

The *De Aspectibus* on the other hand is much more important. In this book Alkindi dealt very explicitly with the optical problem. While he asserted that vision had to take place by means of rays capable of having a physical action upon the eye, he attacked the theory of visual rays which were only mathematical abstract entities and as such were incapable of acting physically and physiologically. According to Alkindi anything which was to act upon the eye had to be three-dimensional and visual rays had only one dimension.

Alkindi transformed and perfected the idea of a ray. He noted that the formation of shadows produced by bodies when illuminated by *lumina* which entered from a window, led without any doubt to the conclusion that the rays emanating from luminous bodies travelled along rectilinear paths.[1] The step forward is considerable when we compare this concept with that of the visual rays and the rather vague and primitive ideas of Lucretius. But Alkindi did not stop there. He affirmed explicitly that vision took place not by means of simulacra or figures which travelled through the air, but rather by means of rectilinear rays which were capable of acting upon the human eye. This, however, was not enough, it was necessary to explain how physical 'rays' which had length, breadth and depth could act within the eye so as to send to the soul the necessary information which enabled it to reconstruct the physical world from which the 'rays' originated. Alkindi did not make any real and effective contribution towards the solution of this problem, but undoubtedly his intervention was positive and useful in his particular field. This explains why, notwithstanding

[1] Alkindi, *De aspectibus* published A. A. Björnbo, S. Vogl, Leipzig, Teubner, 1912, p. 4

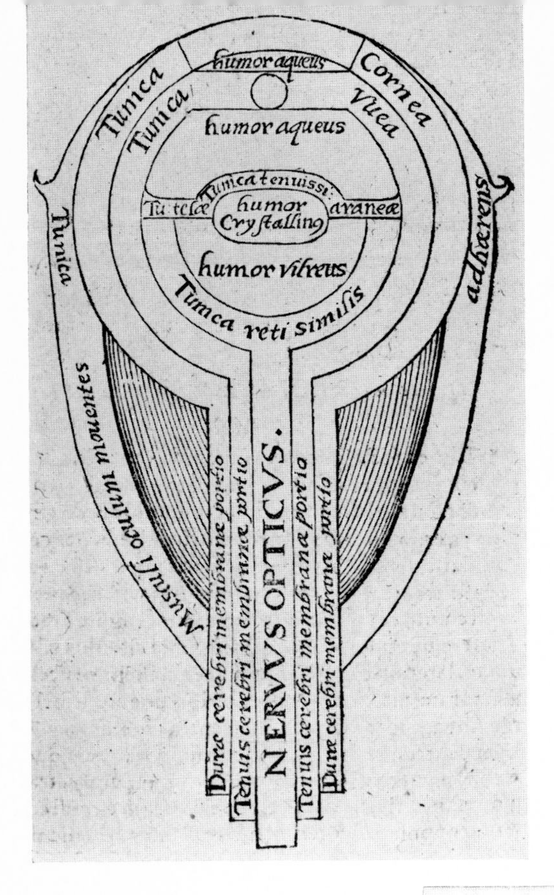

FIG. 1. Diagram of the eye, from *Opticae Thesaurus* by Alhazen

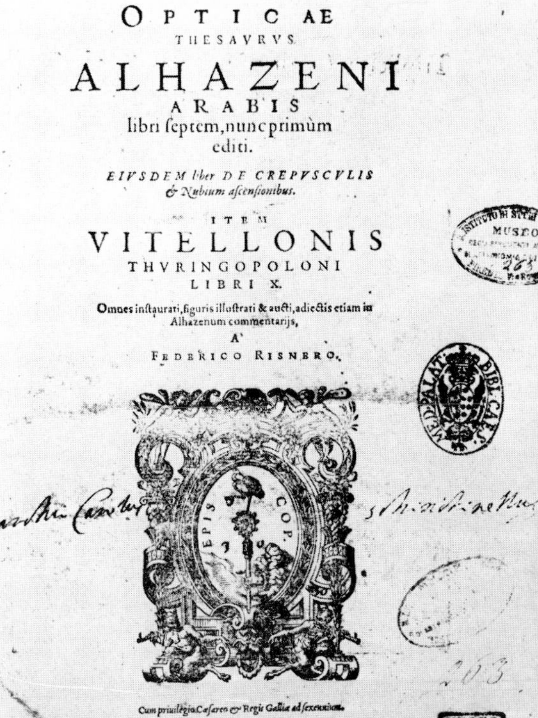

OPTICAE
THESAVRVS
ALHAZENI
ARABIS
libri septem, nunc primùm
editi.

EIVSDEM liber DE CREPVSCVLIS & Nubium ascensionibus.

ITEM
VITELLONIS
THVRINGOPOLONI
LIBRI X.

Omnes instaurati, figuris illustrati & aucti, adiectis etiam in Alhazenum commentarijs,
A'
FEDERICO RISNERO.

EPIS COP.

Cum priuilegio Cæsareo & Regis Galliæ ad sexennium.

BASILEAE,
PER EPISCOPIOS. M D LXXII.

FIG. 2. Frontispiece of Risner's translation of the *Opticae Thesaurus* by Alhazen

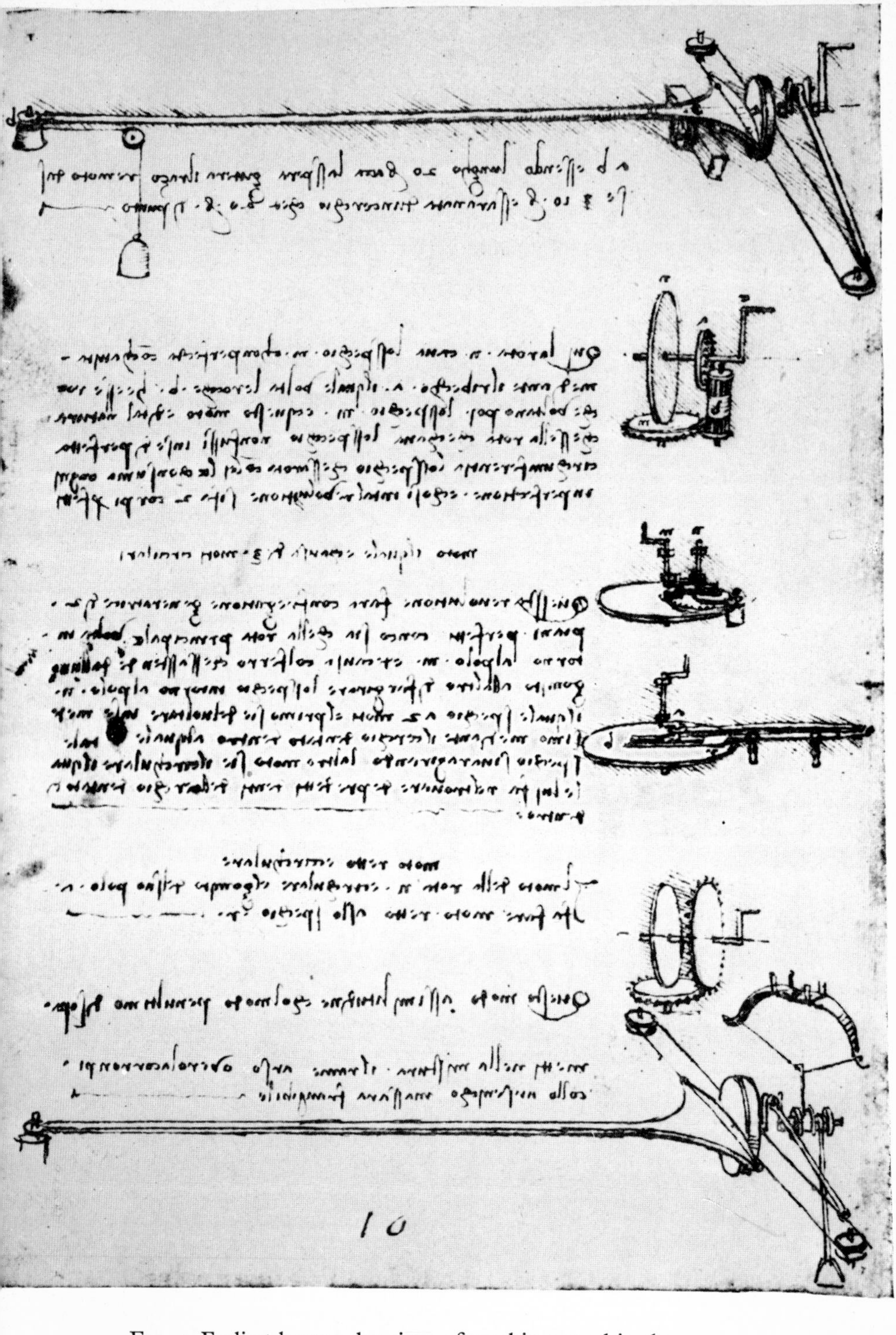

FIG. 3. Earliest known drawings of machines used in the preparation of concave mirrors, from a manuscript by Leonardo da Vinci

FIG. 4. G. B. Della Porta

ALIAE SPECVLI CONCAVI OPERATIONES. *Cap. VI.*

RIVS quàm ab eiusmodi speculi operationibus discedamus, quendam enarrabimus vsum, non parum iucundum, & admirabilem, ex quo maxima Naturæ secreta nobis illucescere possunt. Veluti

Vt omnia in tenebris conspicias, quæ foris à Sole illustrantur cum suis coloribus.

Cubiculi fenestras omnes claudat oportet, proderitq; si spiramenta quoq; obturentur, ne lumen aliquod intro irrumpens, omne destruat: vnam tantum terebrato, & foramen palmare aperito palmaris longitudinis, & latitudinis, supra tabellam plumbeam, vel æneam accommodabis, & glutinabis, papyri soliditatis, in cuius medio foramen aperies, circulare digiti minimi magnitudine, è regione parietes albos, vel papyrum, vel alba lintea appones. Sic à Sole foris illustrata omnia, & deambulantes per plateas, vti Antipodes spectabis, quæq; dextra sinistra, commutataq; omnia videbuntur, & quò longiùs à foramine distabunt, tanto maiorem sibi adsciscunt formã. Si papyrũ, vel albã tabulã appropinquabis, ea visuntur minora, clarioraq;. Aliquantisper tamen immorando, non enim illicò simulachra apparebunt: quia simile validum maximam cum sensu nonnunquã efficere sensationẽ, taleq; inuehit affectionẽ, vt non solum quũ sensus agunt, sensorijs insint, eaq; lacessant, sed etiã quũ ex operibus discessere, diutiùs immorentur, q liquido potest prospici: nã per Solem deambulantes, si ad tenebras conuertimur, comitatur nos affectio ea: vt nil, vel ægerrimè cernamus, quũ adhuc in oculis seruetur affectio ipsa à lumine facta, inde paulatim euanescente, clarè in tenebris aspicimus. Nũc autẽ enunciabo quod adhuc semper tacui, tacendũq; putaui. Si crystallinã lentẽ foramini apponas, iã iam oĩa clariora cernes, vultus hominũ deambulatiũ, colores, vestes, actus, & omnia, ac si propriùs spectares, videbis, tã maxima iucũditate, vt qui viderint, nec vnquàm satis mirari possint. At si vis

Minora omnia, & clariora videre.

E regione speculum apponito, quod non disgregando, sed colligendo vniat, tàm accedendo, recedendoq;, quousq; ad suam veræ imaginis quantitatem cognoueris, debita centri appropinquatione: & attentius cognoscet inspectator volantes volucres, cœlum nubibus dispersum, cyanei coloris, longè distantes montes, & in paruo papyri circulo (qui supra foramen accommodetur) quasi compendiosum orbem videbis, quod vbi vides, non parum lætaberis: obuersa omnia, quia speculi centra vicina sunt, si extra elongabis, maiora, & erecta, vti sunt, conspiciet, sed non perspicua. Hinc euenit

Vt quisq; picturæ ignarus rei alicuius, vel hominis effigiem delineare possit.

Dummodo solùm colores assimilare discat. Hoc non paruifaciendum artificium. Feriat Sol fenestrã, & ibi circa foramen imagines, vel homines adsint, quorum imagines delineare volumus, Sol imagines illustret, non verò foramen. Oppones foramini papyrum albam, ac tandiu homines ad lumen accommodabis, appropinquabis, elongabis, dum perfectam imaginem Sol in obiectam tabulam referat, picturæ gnarus colores superponendo vbi sunt in tabula, & ora vultus circumscribet, sic amota imagine, remanebit impressio in tabula, & in superficie, vt imago in speculo spectabitur. Si vis

FIG. 5. The passage from the *Magia Naturalis* of 1589 describing the camera obscura with lens

DE CATOPTRICIS IMAGINIBUS. 267

Nunc pro coronide adnectam, quod magnatibus, ingeniosis, & studiosis nil fuerit visu iucundius. Vt in obscuro cubiculo obiectis albis linteis venatus, symposia, hostiles acies, ludos, & omnia quæ volueris, ita clarè, perspicuè, & affabrè videri contingat, ac si prę oculis essent. Sit è regione cubiculi, vbi id demonstrare conaberis, planum aliquod spatiosum, quod possit liberè à Sole illustrari, in eo ex ordine arbores accommodabis, sic syluas, montes, & flumina, sic animalia vera, vel arte cosificta ex ligno, aliaue materie, intus pueri consuti sint, vt in comediarum actibus interponere solemus, ceruos, apros, rhinocerotes, elephâtes, leones, & alia quæ volueris animalia effinges: inde paulatim è latibulis egredientia in planum apparent, accedet venator cum venabulis, retibus, alijsq; necessarijs, & venationem simulat, adsint sonitus buccinarum, tubarum, & cornuum; qui enim in cubiculo adsunt, arbores, animalia, venatorum vultus, & reliqua conspicient, vt nesciant an vera, an præstigia sint. Euaginati enses intro per foramen lumen iaculantur, vt ferè terrorem incutiant. Admirantibus amicis multoties eiusmodi spectaculum præbuimus, taliq; illusione gaudentibus, quos naturalibus rationibus, & optices vix ab eorum opinionibus remouere valuimus, etiam artificio aperto. Hinc Philosophis & opticis patet, quo nam fiat visio loco, ac intromittendi dirimitur quæstio, sic antiquitùs exagitata, nec alio vtrûq; artificio demonstrare poterit. Intromittitur idolum per pupillam, fenestræ foraminis instar, vicemq; obtinet tabulæ crystallinæ sphærę portio in medio oculi locata, quod scio ingeniosis maximè placiturum. In nostris in opticis fusius declaratum est. Hinc rei conscio quispiam occultè narrandi auspicari poterit principia, quæ voluerit, vt remotè carceribus occluso. Nec leues poterunt imaginari technæ, distantiam speculi magnitudine emendabis. Sat habes, qui id docere conati sunt, non nisi nugas protulere, nec aliquibus ad huc compertum putarim. Si scire aues

Quomodo Solis eclipsis videri possit

Nunc apponere decreui modũ, quo Solis eclipsis clarè notari possit. In Solis eclipsi claude cubiculi fenestras, atq; oppones foramini papyrum, & videbis Solem, speculo cõcauo in oppositum papyrum resiliat, & circulum suæ rotunditatis describas, sic initio, medio, & fine facies. Vnde sine visus læsione diametri puncta Solis defectus notabis.

FIG. 6. The passage from the *Magia Naturalis* of 1589 describing the similarity between the eye and the camera obscura

Lente crystallina longinqua proxima videre,

Posito enim oculo in eius centro retro lentem, remotam rem conspicator, nam quæ remota fuerint, adeò propinqua videbis, vt quasi ea manu tangere videaris, vestes, colores, hominum vultus, vt valdè remotos cognoscas amicos. Idem erit

Lente crystallina epistolam remotam legere.

Nam si eodem loco oculum apposueris, & in debita distantia epistola fuerit, literas adeò magnas videbis, vt perspicuè legas. Sed si lentem inclinabis, vt per obliquam epistolam inspicias, literas satis maiusculas videbis, vt etiam per viginti passus remotas leges. Et si lentes multiplicare noueris, non vereor quin per centum passus minimam literam conspiceris, vt ex vna in alteram maiores reddantur characteres: debilis visus ex visus qualitate specillis vtatur. Qui id rectè sciuerit accommodare, nõ paruum nanciscetur secretum. Possumus

Lente crystallina idem perfectiùs efficere.

Concauæ lentes, quæ longè sunt clarissimè cernere faciunt, conuexę propinqua; vnde ex visus commoditate his frui poteris. Concauo longè parua vides, sed perspicua, cõuexo propinqua maiora sed, turbida, si vtrunq, rectè componere noueris, & longinqua, & proxima maiora & clara videbis. Non parum multis amicis auxilij præstitimus, qui & longinqua obsoleta, proxima turbida conspiciebant, vt omnia perfectissimè contuissent. Si cordi fuerit

Lente crystallina conuexa imaginem in aere pendulam videre,

Si retro lentem spectabile opposueris, vt per centrum transeat, sed in opposito oculos, intra lentem & oculos spectrum videbis, quod si papyrum obijcies, clare videbis, vt candela accensa supra papyrum ardere uideatur. Sed

Lente concaua quam lata & longa sunt compendiosè depingere.

Poterit pictor maxima commoditate & proportione, nam opposita lentè concaua, quæ in maxima planitie fuerint, in paruum orbem contrahit. Vnde pictor, qui ea contemplatur, paruo labore, & scientia, omnia ad amussim proportionata depingit. Sed ne aliquid de specillis omittamus, trademus

Quomodo res multiplicata videri possit.

Inter ludos, qui circumferuntur, non parum iucunditatis est specillum, instrumentum uitreum illud, quod

FIG. 7. The passage from the *Magia Naturalis* which has been interpreted as a description of the construction of the telescope

the strong attack by Alkindi, the theory of the visual rays continued to be widely accepted and discussed by many scientists for centuries.

Soon after Alkindi, a very great step forward was taken by one of the best known physicists of the Arab School in the transition period between the tenth and eleventh centuries A.D., namely by Abu Ali Mohammed Ibn Al Hasan Ibn Al Haytham, better known in the western world under the name of Alhazen who was born in Basra, probably in A.D. 965, spent most of his life in Egypt and died in Cairo in A.D. 1039.

Like Alkindi, Alhazen wrote many books and was interested in many problems. As far as our subject is concerned the interest lies in a volume consisting of seven Books which in the twelfth century Latin translation was called *De Aspectibus*. The passages to which we shall refer are taken from the translation which, thanks to the interest of Peter Ramus, was published in 1572 by Frederick Risner in Bale under the title of *Opticae Thesaurus libri septem, per Episcopios*. The translation, the first printed edition of this book, was made directly from two copies of the original Arabic text. (Fig. 2.)

Alhazen's contribution to optics has been the subject of many studies some of which are very recent. Among the best known are those made by Wiedemann and by Sarton. When studying Alhazen we detect the influence of the Epicurean School which are shown in *De Rerum Natura* by Lucretius and which underwent a considerable evolution in the work of Alkindi. The exposition by Alhazen however, is very orderly and logical although much of the orderliness is probably due to Risner. We shall consider therefore the exposition systematically in its essential parts, following the logical thread evolved by the author.

The trend of the study is very clearly physico-physiological. The first proposition of Book I states: 'Direct light and illuminated colour strike the eyes'.[2] Alhazen therefore began by siding whole-

[2] Alhazen, *Opticae Thesaurus libri septem, nunc primum editi a Federico Risnerio, Basileae, per Episcopios* 1572. '*Lux per se et color illuminatus feriunt oculos.*'

heartedly with those who admitted the existence of a physical agent, that was external to the sentient individual. In other words he followed in Alkindi's footsteps. From this very first proposition the concept implied in the word *feriunt* is obvious. This concept recalls the fact that the eyes can be damaged by the action of the direct light of the sun as Lucretius had already noted. In fact, Alhazen proved this fundamental thesis with the following reasoning:

> We have found that the eye when looking at a strong light feels pain and may be damaged. As an example, when an observer looks at the Sun he cannot see it because the eye is hurt by the light. The same thing happens if the observer looks at a mirror directed to the Sun, with his eye in the beam of sunlight reflected by the mirror. He will feel a pain on account of the reflected light of the Sun reaching the eye from the mirror, to such an extent that he will not be able to open the eye to look at that light.

If the eye feels pain when it looks at a very bright object it is obvious that we cannot maintain that there is 'something' which travels from the eye towards the object because in such a case there would be no reason to feel pain when looking at one object rather than another. Thus it is necessary to admit that there is 'something' which goes from the object to the eye and this 'something' has the property of also being able to be reflected by a mirror.

There was also another series of observations that strengthened Alhazen's belief in his theory. One experiment was of physiological nature. If we look at a very bright white object *mundum* and we stare at it and then turn our gaze somewhere else, the eye seems blinded and will only gradually regain its faculty of seeing other objects. Alhazen described other similar experiments that produced the same effect using daylight instead of direct light from the sun and even light from a fire. Nowadays we would say that these are effects of dazzle and of persistence of images on the retina. Then followed a remarkable experiment which showed the intention of Alhazen to separate the concept of light from that of a luminous or illuminated body. An observer placed in a room, stares at an opening in the ceiling illuminated by daylight and showing the sky. When he turns his gaze towards a dark part of the room he will still see the shape of the opening but on a dark background.

Alhazen added a very important remark: '. . . and if he shuts his eyes he will still see that shape'. Here Alhazen exaggerates but we can forgive him. He claimed that 'he sees the light' and he did not seem to realize that after all what the observer saw was the diffusing bodies as a whole, similar to the dust in suspension in air seen in a beam of light. Rather than calling attention to his mistake we ought to point out the progressive order of his experiments. He tried to prove the existence of an impression on the eye originating in something external to it. In this case it was the light which entered a darkened room through a hole in the ceiling which revealed the sky but no other object. Even so the eye still received an impression that persisted for a short time, just as when it looked at a well-defined illuminated body. We must note this new observation: when the eyes are closed the impression still persists! Alhazen concluded: 'All this therefore proves that light exerts some sort of action upon the eye'. He did not stop at this, but after having studied and experimented with white light, he repeated the experiments with coloured objects and concluded: 'All this proves that illuminated colours act upon the eye'.

We have described at length these experiments because we are dealing with a text which is not very well known and also because to us it seems rather surprising to find in the early eleventh century, a physicist who was following experimental methods almost comparable to those of modern times. We must admit that we have come a long way from the works of Euclid to this first chapter of Alhazen's book. Already in the first pages there are two things which are of great interest. First the controversial tone, since Alhazen puts forward several ideas which were contrary to the general opinion of the time, rather than safely repeating ideas that were generally accepted. Secondly the introduction of a new concept in the word 'disposition' (*dispositio*) of the eye. This concept means that the eye has the power of seeing without the inconvenience of dazzle or of the persistence of the images. Later Alhazen explained what he meant by *dispositio*.

In the second Proposition of the same book, Alhazen related a great number of experiments which we would define nowadays as 'visual sensitometry'. He began by stating that stars could be seen only at night because during the day there was the light of the sky. He performed other experiments using intense fires and objects which had very fine details such as miniatures and fine writing, and proved that objects or their details which could easily be seen, or at least perceived in good lighting, became invisible when illuminated by light that was either too bright or too weak or when the flames were in such a position that the objects were seen 'against the light', as we say. In the third chapter of Book I the writer showed with the support of many observations, that the tone of the colour of objects varied as the light which illuminated them changed. With this first group of experiments and with the accompanying discussion Alhazen wished to eliminate the idea that rays could be emitted by the eye to go out in search of the objects. Then, contrary to his numerous predecessors who had found it easier to destroy old views than to suggest new ones, he proceeded towards a phase of constructive ideas and overcame the main difficulties that confronted all those who before him had thought of a 'something' which from the objects entered the eye. Alhazen soon realized that he had to avoid the idea of 'eidola' and of 'husks' which could not in reality stand up to even the most elementary criticism, and at the same time he had to avoid also the idea of a structureless fluid which could not have given to the eye the information required about the shape, dimensions and position of the object. He succeeded in finding the key to the mystery and started a line of thought which has continued up to our day. It is strange that a man of such merit should be almost forgotten.

In the fourth chapter of his work, Alhazen described the structure of the eye according to the teachings of Galen and of the Arabic medical school. In the next chapter he discussed the mechanism of vision under the title: 'Vision is produced by rays emitted by the objects towards the eye'.[3] The title of this chapter stated in essence the idea which Alkindi had already advanced more than a century earlier and which Alhazen accepted without reservation. Here again we find Euclid's rays but this time they proceed from the object to the eye. The proof begins thus:

[3] Op. Cit. Lib. I, Chap. V, p. 7

It has already been made clear (chapter I) that a body illuminated by any light, emits a light in opposite direction.[4]

Is this not a new optics very different from that of the Graeco-Roman world? Alhazen continued with clear logic:

> When the eye is in front of an object illuminated by any light, some light must reach the surface of the eye. Since we have already shown that light has the property of acting upon the eye, and that it is the nature of the eye to react to light, we must conclude that the eye cannot perceive the object which is seen except by means of the light that the said object sends to the eye.

Eidola and husks are dead. Light travels from a source on to objects and these send it back in all directions and hence also to the eye that is in front of them. This is a considerable step forward, more for the sense of necessity that is implied than for the explanation of the mechanism itself. There still remained the fundamental problem to be solved. How could the light which entered the eye convey to the sensitive organ the form and structure of the object? Alhazen suggested a new idea; if the eye were endowed with sensitivity only without having any directional elements, then the eye would only be able to see lights and colours as mixed and indistinct. The only way to explain distinct vision was to assume that to every point of the observed object there corresponded in the eye a stimulated point. This was a decisive step forward.

At the time of Alhazen, in spite of the existence of the optics of Euclid, Ptolemy and Damianus, the general knowledge about refraction was insufficient to enable anyone to trace, even in a schematic form, the trajectory of light through the transparent membranes of the eye. This situation not only justifies the errors committed by Alhazen in defining his ideas in detail but increases considerably the value of his ideas and gives them a masterly and almost prophetic character. The study of these ideas in detail leads us away from the main subject, nevertheless it is worth considering the general lines followed in the reasoning. Alhazen reverted to an

[4] Op. Cit. p. 7.

idea which had already been advanced by earlier scholars and which Alkindi had shown to be false and unacceptable. The idea was that a ray which fell obliquely on to a transparent surface was subject to refraction, namely that such a ray was bent and as a result of this was greatly weakened. Of all the rays which reached the cornea from a point of an object observed, only one of them was perpendicular to the cornea and therefore went through it without being refracted and only this ray was completely effective. According to the structure of the eye accepted by Alhazen, all its internal membranes were supposed to be concentric and therefore the ray would travel through them all without being refracted. The other rays that entered the pupil were refracted namely broken and made ineffective, even those rays very near to the one that entered the eye at right angles to the surface of the cornea. With this idea it was possible to overcome the difficulty which arose when the source of rays was a point source. The supporters of the theory of 'husks' had met with an unsurmountable difficulty in trying to explain how the husk of a mountain, for example, could enter the pupil that had a diameter as small as 2 millimetres. Now the pupil was really too large, because of all the rays that reached the eye from each point of the object, only one could be used and to each point of the object there corresponded only one point on the sensitive surface of the interior of the eye. We must admit that this concept may appear today somewhat forced although it was very advanced and almost prophetic coming from someone who did not know the concept of optical axis and of correspondence between points of the object and points of the image.

With this approach Alhazen overcame the two main obstacles which, as we remarked at the beginning of this chapter, had stopped all further development of previous theories concerning vision. Not only did he make use of the contributions of Galen and of the medical school of Baghdad in order to overcome the idea that the eye acted only as a centre of perspective, but he also paid attention in great detail to the interior of the eye. Furthermore, he divided the visible object into point-like elements and in this way he suppressed the character of an indivisible and total operation which until then was attributed to the act of seeing the object.

With regard to the functioning of the eye, Alhazen went much deeper than the ideas we have already mentioned. Starting from the general description of the eye given by Galen and assuming that the various transparent layers were concentric and that the crystalline lens, situated at the centre of the eye was the seat of the sense of vision, Alhazen re-examined the concept of the pyramid of perspective of Euclid. He rebuilt it with those rays that started from single points of the object and arrived at right angles to the surface of the eye, and as a consequence he placed the apex of the pyramid at the centre of the eye. Light and colours of the various points of the object had to proceed in an orderly manner along these lines passing through the centre of the eye maintaining the same order also when passing through the cornea and the pupil. They were thus imprinted on the first surface of the crystalline lens which was sensitive to them. The image, similar to the object, had thus entered the eye and had brought in their proper order all the elements necessary to produce vision. Of course there are many errors in these ideas but they have the value of being theoretical ideas which leave the past behind and which will survive.

We may ask why Alhazen knowing the existence of the retina and its nerve structure should suggest that the impression of the image took place on the first surface of the crystalline lens? Obviously because a very considerable difficulty would have been created if he had made the rays travel beyond the crystalline lens and therefore beyond the centre of the eye. The order of the image would have been inverted and on the retina everything would have been upside down. For fifteen centuries there had been the unsurmountable difficulty of explaining how the image of an object as large as a mountain or the whole sky could enter the small pupil of the eye. Alhazen with a stroke of genius surmounted this difficulty but he would have been faced with the collapse of the whole of his edifice if the image was inverted from the natural order. He avoided the collapse of his theories by theoretical juggling. He suggested that the image was perceived before being inverted, namely before it reached the centre of the crystalline lens although there were only completely transparent surfaces surrounding it. This was not the only rescue operation which Alhazen was compelled to undertake. Evidently he felt that in his theory there was something valuable. When he met with difficulties that appeared unsurmountable to him, he put forward temporary solutions even if he did not find them completely satisfactory, hoping that the future would rectify them. This, after all, is the line which has been taken by philosophers the world over. Later Alhazen himself discovered that objects can be seen even if a small obstacle, such as a needle, is placed in front of the pupil so that some of the rays perpendicular to the surface of the eye are intercepted. He also discovered that objects are visible even if they are in such a position that normal rays from their various points cannot reach the cornea: in other words, objects which nowadays we would say are at the edge of the field of view. He explained all this by the refraction of rays which enter the eye at various angles and finally reached the conclusion which pleased him that 'all vision takes place because of refraction'. Nevertheless he did not attempt to explain how the order of the image which entered the eye could be maintained when the principal rays were broken and intercepted or when refraction intervened. More research was required to obtain the basic knowledge necessary to eliminate the difficulties surrounding Alhazen's theories.

Alhazen, after having asserted that vision was not produced by rays emitted by the eye ('*visio non fit radiis a visu emissis*') went on to compromise and he maintained that vision seemed to take place by both rays received and rays emitted. He explained that the existence of rays emitted by the objects towards the eye was not really sufficient but that it was necessary that the eye should be directed towards the object so as to receive the rays and, at the same time, the eye had to have that *dispositio* which enabled it to see. He then specified that the impression of these external rays could give only the sensation of light and colour and in order to complete the process of vision, and to see the object in its place, many factors of psychological nature were required and these factors he summarized as: '*cognitio et distinctio antecedens*' to which he also added '*argumentatio iteranda apud visionem*'. Continuing the analysis of the work of Alhazen we find that he devoted an entire Book, the third of the seven, to optical illusions which he calls *deceptiones visus*. He also described a great number of experiments relative to vision, to the field of view and to the acuity of vision. Unfortunately we cannot discuss all these here because they are not directly connected with our subject. Briefly we must recognize that Alhazen is the true founder of what today we call 'physiological optics'.

Before we end this resumé of the extensive work of this great scientist who lived over a thousand years ago, we must mention his views on the nature and properties of light. On the basis of his theory of the mechanism of vision he built a model of light assuming light to be something which was propagated by linear trajectories and which could stimulate the sensitive part of the eye to such an extent as to hurt it, or at any rate, to render it incapable of seeing for a time that could vary in length. In addition, he attributed to this light the property of reflection when it met a mirror, and of refraction when it travelled through transparent surfaces. This light, according to him, started from luminous objects and was diffused all around through diaphanous bodies. With reference to these bodies he made it quite clear that they had no power to alter the light they received in any way. To prove this he carried out an experiment that can be called the experiment of *camera obscura*. He placed several candles in front of a wall with a hole and looking at a screen placed on the other side of the wall, he saw on it as many images as there were candles. These images were found along lines which crossed at the hole. When he suppressed one of the candles he noticed that the corresponding image also disappeared, and when the candle was replaced the image came back in the same place. All we need to make this experiment agree with the principle of the *camera obscura* is the remark that each image is upside down. Alhazen did not attribute any geometrical importance to this experiment and discussed it only as far as it was of interest to his argument. He simply deduced that if the various lights mixed together in the air then we would be unable to see the images of the candles. He concluded therefore, that light followed straight lines in diaphanous bodies without danger of being mixed with other lights. An interesting question arises at this point. Alhazen described this experiment in his first Book which is the one devoted to the mechanism of vision. He showed that the membranes of the eye were involved in the experiment and so, a little later, he felt compelled to introduce the exception we have already mentioned about the first surface of the crystalline lens. Since the crystalline lens was sensitive and since it could be hurt by light then it had to be considered to be an exceptional transparent body. '*Humor crystallinus lucem et colorem aliter recepit, quam coetera corpora*'. It is remarkable that these conclusions reached by Alhazen follow almost immediately the experiments of the *camera obscura*. This very experiment five or six centuries later will be responsible for the final step towards the explanation of the mechanism of vision. Let us return now to Alhazen's 'light'. If the diaphanous bodies did not alter it, the 'dense' or opaque bodies produced much more complex effects. When opaque bodies receive light from any source, be it the sun, daylight, fire or candles, they send it back in all directions. We should note that the colour of the illuminated bodies changes with the changing of the light which illuminates them. Here we meet with an objection: why is it that when we look at a body in front of our eyes we see both the form and colours of this body and not those of the body which illuminates them? With his usual perspicacity Alhazen remarked that the 'forms' of an object grew much fainter as they travelled away from the object from which they had detached themselves. It was true therefore, that when we looked at an illuminated object we had to receive the forms and colours of both the object and of the source of light, but the prevalence of one or other was purely a question of intensity, since after all the source of light had to be further away from our eyes than the observed object. So in this connection he repeated in a different way, the experiment of the camera obscura. He interposed a screen with a hole between a body illuminated by any light and another white body placed in darkness, or in other words, he built a camera obscura containing a white body in front of the hole, and he observed on this white body, acting as a screen, the light coming from the source without.

In all the work of Alhazen in spite of the evidence we find of a new mental approach and of an effort directed to co-ordinate phenomena studied experimentally, we still discover a certain influence of the older methods of reasoning and more in particular of that of the atomists. The persistent distinction between light and colour, the reference to a 'primary form' which starts from the source and falls upon the illuminated body, and a 'secondary form' which detaches itself from the body and enters the eye, are all still very reminiscent of the *eidola*, even if divided into smaller parts and purified. The past cannot be destroyed so easily after all! On the other hand we must call attention to some ideas which were to be developed in the future. Indeed, we can say that very few people seem to realize that these ideas were put forward in the eleventh century. We are referring in particular to the parallel between light and projectiles.

In Book IV, Alhazen studied reflection upon all kinds of mirrors, plane, concave, convex, spherical, cylindrical, etc. He studied very carefully the formation of images and the curious effects produced, but unfortunately we cannot spend too much time on these questions. Naturally Alhazen followed Euclid's law of reflection, but he pointed out that the rays had to be considered as received rather than emitted by the eye. He wished, however, to give a 'natural' proof of the law of reflection and therefore he made use of an example in mechanics. The text in question says:

> The reason why light is reflected by a mirror at an angle equal to that which the light had on arrival on the surface of the mirror, is the following. Light travels at a very high speed and when it reaches a mirror is not allowed to penetrate nor is it allowed to stop there and, since it still has the force and nature of its original motion, it is reflected on the side from which it came, along a line which has the same inclination as the first line. [It is difficult to realize that this reasoning was put forward nearly a thousand years ago!] We can find a similarity to this in both natural and artificial motion. If we let drop a heavy spherical body from a height on to a perfectly smooth plane, we notice that it is reflected back following the same perpendicular line which it followed in its fall. The same applies in artificial motion. Let a mirror be fixed to a wall at the height of a man and let a small spherical body be fixed at the tip of an arrow. If we then release the arrow by means of a bow from a height equal to that of the mirror and directed at right angles to the mirror, we shall see that the arrow will bounce back along the same perpendicular line. On the other hand if the arrow is tilted at an angle to the perpendicular we shall see that

> the arrow is reflected back not on the same line by which it came, but in a direction which is symmetrical to the first with reference to the perpendicular to the mirror.

Alhazen gave an explanation of this behaviour by resolving the motion into two components, one normal and the other parallel to the plane of the mirror, the first being inverted by reflection and the other component being unchanged. He explained this for heavy bodies and then he added:

> Light has the same nature of being reflected, but does not have the property of ascending or descending. Therefore during reflection it moves following the straight line on which it originally started until it meets the obstacle that will stop the movement... This is the cause of reflection.

The part which refers to refraction is equally interesting. Alhazen again resolved the motion into two components, one normal and the other parallel to the transparent surface, and while he left the second unchanged, he considered the first as accelerated or retarded. The examples from mechanics he used are as follows: if a small iron ball is thrown towards a very thin wooden screen so as to make a hole in it, we can see that the trajectory after having perforated the screen is nearer to the normal. Similarly if someone tries to cut a wooden plank obliquely with a sword he will see a similar phenomenon. These examples are interesting but Alhazen's reasoning to justify the variations in the normal component of motion is quite remarkable.

> Lights which travel through transparent bodies travel with rapid motion which cannot be appreciated because of its velocity. Nevertheless the motion through thinner bodies, that is to say diaphanous bodies, has a greater velocity than in thicker bodies which are less diaphanous. In fact every diaphanous body opposes a certain amount of resistance to the light travelling through it and this resistance depends on the thickness of the body.

As we shall see later, the situation is not quite like this because when the normal component is braked the ray moves further away from the normal instead of getting closer to it. Nevertheless in this 'light' of Alhazen we sense the existence of an idea which we tend to attribute to a more modern age. How did it happen that so many new ideas, so well co-ordinated and based on experimental methods were ignored for so many years? The explanation can be found in the fact that Alhazen lived in the Arab world at a time when the western world was passing through a very difficult period of its history. We shall see later what were the obstacles in the realm of philosophy which made the penetration of these new ideas in the west so difficult, even in the centuries immediately following the tenth century.

There exists a treatise on optics by Witelo, a man about whom we know very little except that he was of Polish origin and that in 1271 he was at Viterbo in Italy. This treatise was studied very closely in the Middle Ages, first in manuscript form and then as a book in 1533. It is a book which from the formal and purely mathematical point of view is much better than that of Alhazen but as far as the physics of light is concerned, it only repeats the views, the proofs and the experiments of Alhazen, without acknowledging the source.

When in 1572 Frederick Risner translated the original Arabic text of Alhazen, he stated in his preface that he had not altered the original text but had only improved the order of exposition and the arrangement of the seven Books. In this edition he followed Alhazen's text with the *Opticae Libri X* by Witelo and to each Proposition of this book he added useful notes to show which passages of the Alhazen's text Witelo had used to prove his points. Thus very slowly the ideas of the Arab School reached the west and permeated its cultural centres, which at the time were beginning to recover from the Dark Ages. The ideas of the Arab School were based on physiology and on physics and also took into account the very important part that psychology played in the complex phenomenon of vision.

During the Greek period which we studied in the previous chapter, the supporters of the theory of 'husks' which had a physico-physiological basis were in a small minority when compared with the supporters of the theory of the visual rays, which was basically of a mathematical character. Likewise in the eleventh century the followers of the ideas propounded by the Arab School and which were of a physico-physiological character were indeed a tiny minority against a very powerful new group of philosophers. We can say that at this stage the minority supporting the theories based on physics, had succeeded in defeating the supporters of the mathematical theories, notwithstanding their very great number, and had compelled them to find refuge in the study of purely abstract theories such as those of geometrical optics and of perspective. Hardly had the minority supporting the physical theories surmounted this obstacle than they were faced with another equally difficult obstacle. We have reached now that period in history when culture in the west was beginning to re-awake, when the ancient manuscripts were rediscovered and when the diffusion of the classical texts carefully translated was taking place. All this was occurring mainly in the ecclesiastical centres and in some centres of knowledge such as Bologna, Paris and Montpellier which were more or less under the control of political and religious authorities. Apart from the external influences and the directions from the authorities it was natural, indeed inevitable, that all the scholars who approached the learning of the golden Graeco-Roman period were bound to find it even greater and more wonderful than it really was. We must not be surprised therefore, if by a natural and spontaneous process, a kind of dogmatism arose. Tradition and the authority of the great masters were arguments superior to all others even when these were supported by the greatest experimental evidence. On the other hand, because of the political climate of the times, this type of mentality was widely supported and encouraged by external factors which controlled the cultural atmosphere of the age. It is well known that in those days, when confronted with any phenomena rather than study them by observations and experiments, it was customary to search for an explanation in the classical texts. Higher education and the training of new teachers consisted mainly in teaching how to interpret and how to understand the 'unrivalled' classical texts. In such an atmosphere, ruled by a mentality which was conservative to excess and paralysed by a sincere admiration for the greatness of ancient learning, it was natural that the experimental physicists should be considered as revolutionaries and were therefore boycotted by the heads of the cultural centres. It was inevitable that only very few felt that they could face a situation so difficult and so costly, because even in those days the possibility of a brilliant career was dependent on the ideology that a man held.

If this were true of the West what was happening meanwhile in the East? We must not think that the scientific world was dominated by personalities like Alkindi and Alhazen. The Arabic medical school was encouraged and supported because of the practical and important necessity of curing illness. Much more famous than Alhazen was his contemporary Abu Ali Hosayn ibn Abdullah ibn Sina, known in the West under the name of Avicenna (980–1037). He became one of the best known of the mediaeval philosophers and his influence spread also to the West. Of all his writings which are very extensive, we are interested only in his *De anima o Liber sextus naturalium*. He studied with great care and in detail the problem of knowledge of the external world, and therefore the mechanism of the senses and in particular the functioning of the sense of vision. His ideas however, followed a very different path from that of Alhazen. The latter explicitly recognized that in the process of vision there was a very important intervention of the mind (psyche) of the observer that played a very important part in the total process, at the same time he also realized that the first phase was purely physical and that this was followed by an intermediate physiological phase (if we wish to use modern terminology). While Alhazen attempted to define as best he could the structure of the physical stimulus and the effect of the physiological intervention, Avicenna minimized completely the physico-physiological part and concentrated his attention on the psychological activity of the observer. He considered this activity to be essential and most important, while the sensorial part, according to him, was only occasional and without importance. From this it is a very short step to doubt the existence of the material world. We can say therefore, that Avicenna is not only an anti-materialist but he is also against the idea which was supported by Alkindi and Alhazen, that the rays capable of acting on the eye were of a material nature. Avicenna re-introduces the theory of emission of simulacra from bodies, but strips these simulacra of all material form and ends by calling them 'forms' or *species*. These ideas were very well received in the West and received much more support than the ideas of the materialists. Avicenna was considered one of the greatest 'masters' and his ideas found ready response in the philosophical mentality of the West. In fact many of the scholars of the West who were studying the problem of vision were strongly influenced by Avicenna.

Eleven years before Avicenna died, Averroes (1126–1198) was born. Although he was born in Cordova, in Spain, he belongs to the Islamic culture. His full name was Abul Walid Mohammed Ibn Ahmad Ibn Mohammed Ibn Roshd and he was another famous philosopher who had a great influence on the western world. He took an opposite view from that of Avicenna and he met with great opposition in official quarters, even in the Islamic world, so much so that his writings were burnt by royal decree. This shows how strong and how deeply felt were the polemics on fundamental philosophical ideas. From the point of view of our history of light it is important to note that Averroe openly challenged Avicenna's ideas concerning the mechanism of vision but he kept the argument purely in the philosophical field, without discussing the physico-physiological question of which Alhazen was still the most important exponent.

If we wish to summarize the theories about light prevailing during the last centuries of the Middle Ages, we can say that discussions and investigations revolved around three ideas, namely in the field of mathematics, physics and philosophy. The first had lost considerable ground and restricted itself to abstract studies concerning geometrical optics and perspective. The other two were holding their ground vying with each other for the importance of their theories. In effect at this time a division was taking place not only in the direction of thought but also in terminology. The followers of the physical theories supported the idea of the existence of an external agent which was capable of stimulating the sense of vision and they finished by adopting the term *lumen* to indicate this par-

IOAN. BAPTISTAE PORTAE NEAP.

DE REFRACTIONE OPTICES PARTE.

Libri Nouem.

1 *De refractione, & eius accidentibus.*
2 *De pilæ cryſtallinæ refractione.*
3 *De oculorum partium anatome, & earum muniis.*
4 *De viſione.*
5 *De viſionis accidentibus.*
6 *Cur binis oculis rem vnam cernamus.*
7 *De his, quæ intra oculum fiunt, & foris exiſtimantur.*
8 *De Specillis.*
9 *De coloribus ex refractione, ſ. de iride, lacteo circulo, &c.*

Ex Officina Horatij Saluiani.

NEAPOLI, Apud Io. Iacobum Carlinum, & Antonium Pacem. 1593.

FIG. 8. Frontispiece of *De Refractione* by Della Porta. Ed. 1593

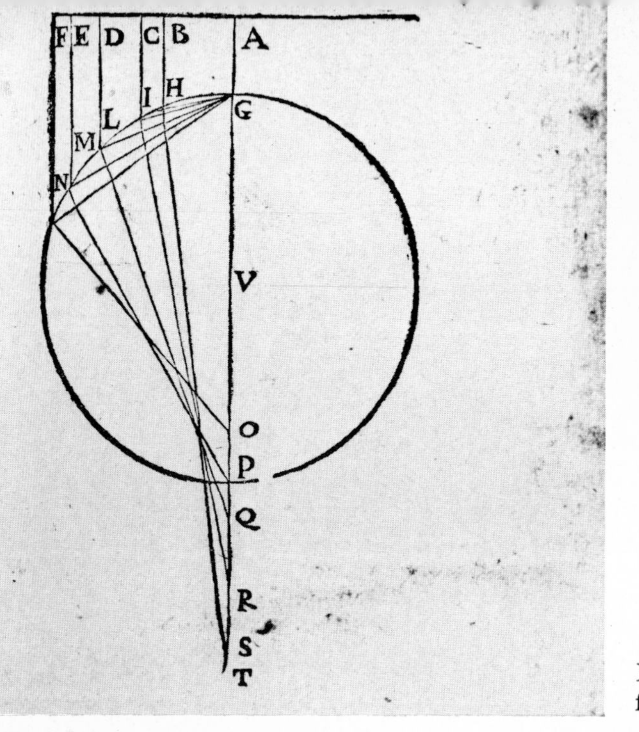

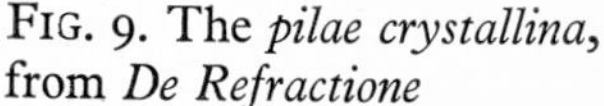
FIG. 9. The *pilae crystallina*, from *De Refractione*

DE PILAE CRYST. REF. LIB. II. 47

Vera loca determinare Solaris radij refractionis egredientis ex conuexa sphærali superficie. Prop. V.

HI anguli sunt æquales illis, qui ingrediebantur in superficiem vitream conuexam, attamen contrarij, nam radij, illi Solares in contrariam partem labuntur, vt videbimus in subiecto exemplo.

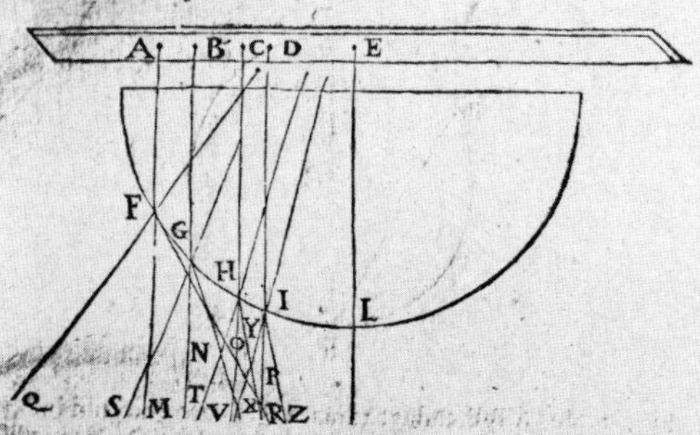

Esto tabella Soli opposita, & perforata A B C D E, semicirculus vitreus Soli oppositus F G H I L, cuius recta diameter Soli obuersa, circumferentia aduersa.

Pergat radius E L, transiens per semicirculi centrum, penetrat irrefractus.

Deinde A F tangens F punctum, latus exagoni à puncto L, non ab A ad F penetrat perpendiculariter, & irrefractus, deinde foras egrediens ex F, in contrariam partem labitur in R, quãdo enim superficiem tangebat connexam, veniebat per F Q, rectus radius F M, nunc labitur in R, & angulus Q F M, erit æqualis M F R.

Sic radius C H, tangit latus duodecagoni, in superficie conuexa veniebat per H T, nunc per H Y, & angulus D H O, æqualis est angulo O H Y.

Sic radius B G tangit latus octogoni, labitur in X, veniebat G S, & D I, latus sexdecagoni labitur in Z, veniebat enim per I V.

Dato

FIG. 10. The *convexa sphaeralis superfices*, from *De Refractione*

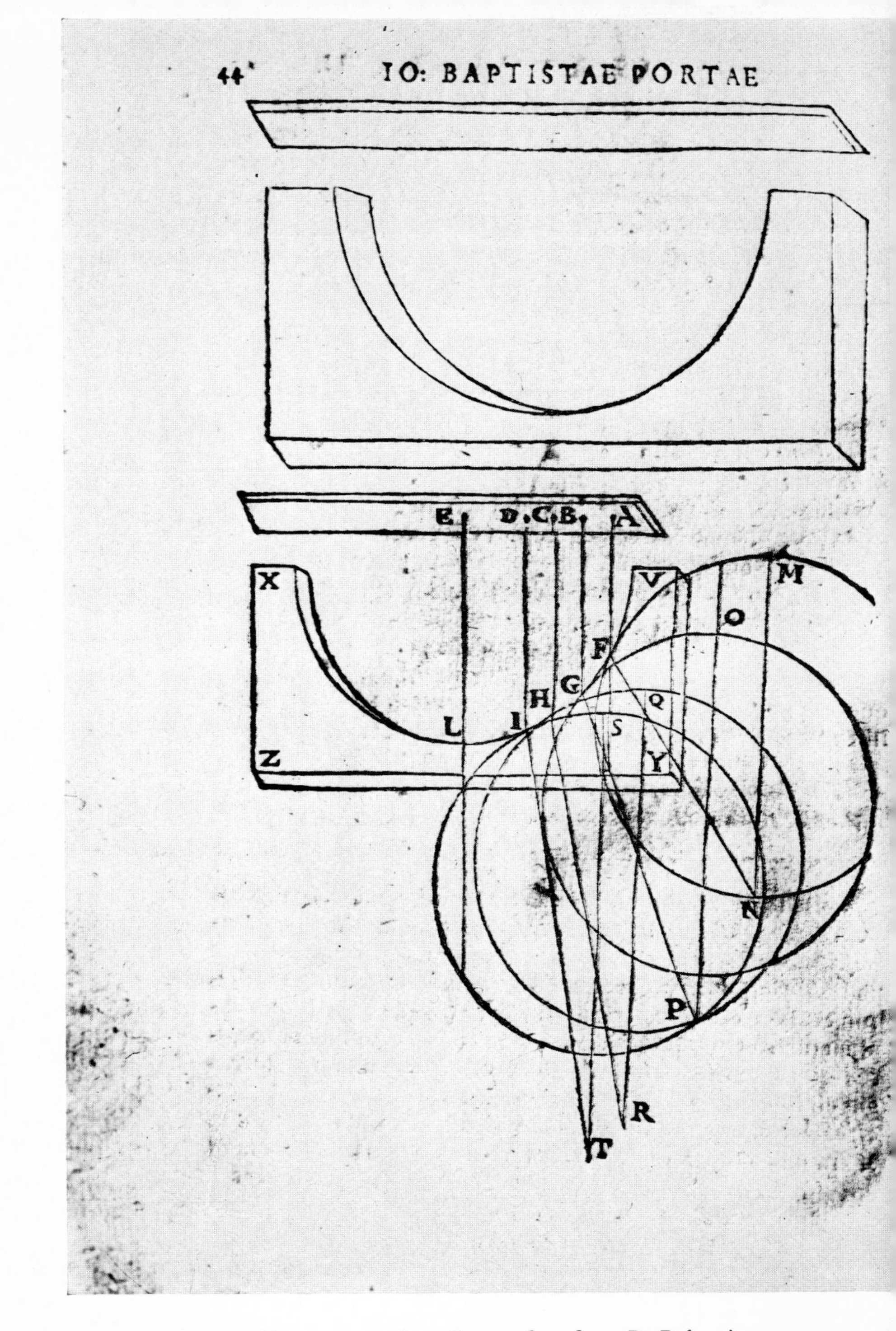

FIG. 11. The *concava sphaeralis superfices*, from *De Refractione*

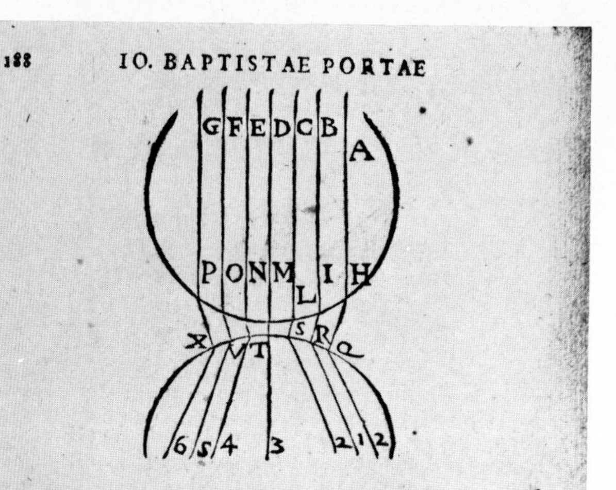

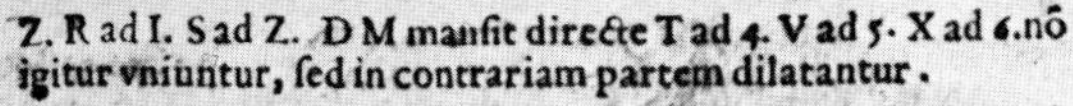
Z. R ad I. S ad Z. D M mansit directe T ad 4. V ad 5. X ad 6. nō igitur vniuntur, sed in contrariam partem dilatantur.

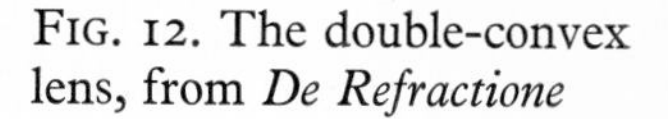
FIG. 12. The double-convex lens, from *De Refractione*

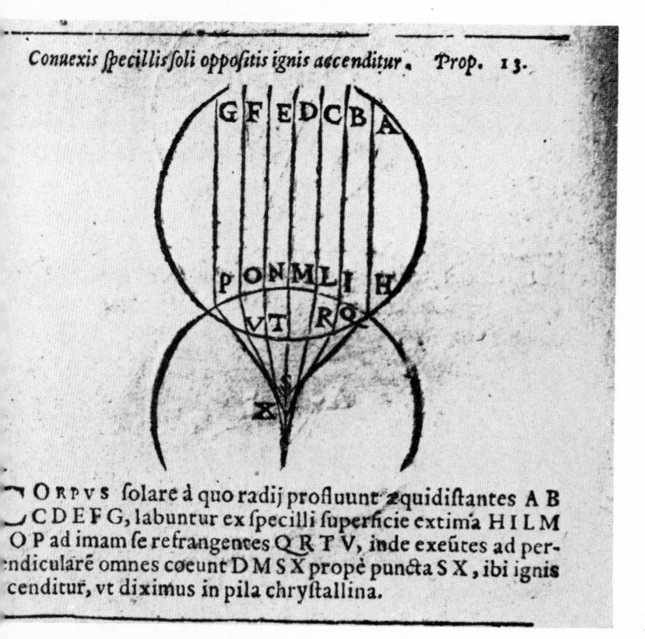
Connexis specillis soli oppositis ignis accenditur. *Prop.* 13.

ORPVS solare à quo radij profluunt æquidistantes A B C D E F G, labuntur ex specilli superficie extima H I L M O P ad imam se refrangentes Q R T V, inde exeũtes ad per-ndicularẽ omnes cœunt D M S X propè puncta S X, ibi ignis cenditur, vt diximus in pila chrystallina.

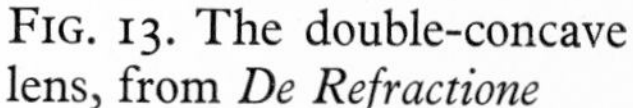
FIG. 13. The double-concave lens, from *De Refractione*

ticular agent. The followers of Avicenna on the other hand, minimized the importance of *lumen* and tended to concentrate their interest on the spiritual representation or, in other words, on 'brightness' which the observer sees when he states that it is not dark. To indicate this entity the word *lux* was adopted. It was never explicitly stated that *lux* was a subjective 'psychic' entity or that it existed together with colour, only in the mind of the observer. Without committing themselves the supporters of this view gradually considered that *lux*, like shape was something existing on the surface of bodies. Since the shape of a body, according to classical ideas, revealed itself to an observer by means of simulacra which by now were more frequently called *species*, the idea took root that *lux* revealed itself to an observer by means of *lumen*, or, in other words, *lumen* was considered as the *species* of *lux*.[5] Thus while physicists devoted themselves especially to the study of *lumen* and tended mostly to attribute to it a material nature, according to the atomistic theories, philosophers, on the other hand, attempted to define the nature of *lux* with a kind of reasoning that could almost be called the 'metaphysics of light' with incursions even into the theological field. Since most of the works on this subject were written in scientific mediaeval Latin, they were given the title of *perspectiva* which should be interpreted as meaning the 'science of vision'. The word *perspectiva* was taken as the translation of the word 'optics' which was used by the Greeks for the same purpose, using the same root which in the Greek language referred to 'seeing'. The mediaeval *perspectiva* is only faintly related to our

[5] In tracing the evolution of the mediaeval ideas of light we have maintained the Latin terminology because we have to be extremely careful when we attempt to translate these terms in modern languages. *Lumen* and *lux* were translated in the Italian mediaeval scientific texts by *lume* and *luce* with the same meaning as the corresponding Latin terms. In this sense these two words were used by Dante, by Leonardo da Vinci by Galileo and by all writers who wrote in Italian on scientific matters until the seventeenth century. This particular question will be discussed again later (see page 224). In order to express as faithfully as possible the ideas of the various scholars we are going to discuss, it is essential that we should use the two words *lumen* and *lux*.

modern 'perspective' which in reality is only a derivative of it. Our modern perspective is rather linked to the ancient theory of the Greek mathematicians, namely to the theory of visual rays.

The greatest exponents of the mediaeval *perspectiva* flourished in the thirteenth and fourteenth centuries. The first of these was Robert Grosseteste (1175–1253) Bishop of Lincoln. He was followed by Roger Bacon (1214–1294), by John of Peckham (1220–1292) Archbishop of Canterbury, by Saint Bonaventura (1221–1274) and St. Thomas Aquinas (1227–1274) who were all contemporaries and more or less of the same age. Another Franciscan of importance in the thirteenth century was Bartholomew of Bologna who lived a few years later than the above mentioned. Of particular interest is the treatise written by Robert Grosseteste entitled *De iride seu de iride et speculo*, two manuscripts of which are in the National Library in Florence. Roger Bacon dealt with the subject of the nature of light and other optical questions in the fifth part of his well known *Opus maius*. John of Peckham wrote *Perspectivae Communis Libri tres* which although a comparatively small book was one of the most well-known books during the late part of the Middle Ages. St. Bonaventura dealt with the question of vision particularly in the second Book of *Commentarium in quattuor libros sententiarum* and St. Thomas Aquinas discussed the same question in various parts of his many works. Bartholomew of Bologna was the author of *Tractatus de luce*. Of course the above books were not the only works published on this particular subject during the thirteenth century, but we mentioned these to show how much thought and work was devoted to the problem of vision during this period of history. It is also interesting to note that all these writers, and no doubt they were the most important, they all occupied very high positions in the Church. There is not much point in examining in detail the differences between their views, especially since their discussions had a tendency to be more concerned with metaphysics and the nature of knowledge than with technical or scientific aspects. The earlier writers were strongly influenced by their predecessors and in particular by St. Augustine and by Avicenna. Although at first the theological tendency was so strong as to lead even to the identification of *lux* with God, later *lux* began to be considered as a divine emanation and finally it was argued whether *lux* was a material or spiritual entity. The language is very difficult to follow because the terms used were, naturally, those used in the context of the philosophy of the time. The definitions of 'ray', *species* and 'form' were very complicated. The exposition of the theory of knowledge, of which vision was considered to be a particular but important case, was very complex and not always understandable. In all the writers there is evidence of the acceptance of the ideas of Avicenna in full, even if there were a gradual tendency to lessen the importance of intuition and of rational intervention and to take into account the contribution made by the knowledge obtained through the senses. In this way a model of vision was outlined that was a combination of the action produced on the eye by the *species* and the receptive capability of the psyche of the observer. The psyche of the observer however, was considered to predominate. On the question of the nature of *species* a very lively argument developed between those who were prepared to attribute to them a structure that was material even though extremely delicate, and those who considered them purely as a spiritual entity. Of course 'rays' were also mentioned but never in the sense adopted by Alkindi and Alhazen: indeed they were often considered as synonymous with *species*. In the philosophical discussions the 'visual rays' were almost completely forgotten even if mention was sometimes made of the pyramid of perspective. Although even Roger Bacon tended to follow this line, nevertheless we can detect in his work an attempt to encourage experimental activities. In this sense the experiments which he carried out and described in the field of reflection and refraction of *lumen* are noteworthy. Indeed we can safely say that among all the thinkers of the time, Bacon is the one whose line of thought is the nearest to that of Alhazen.

To conclude this brief summary of the ideas on *perspectiva* of the thirteenth century, we shall quote a passage taken from chapter IX of Treatise III of the *Convivio* by Dante Alighieri:

> ... that what is really visible is colour and light as Aristotle tells us in the second book of *De anima* and in the book on *Sense and Sensation*. A

visible thing is something quite different because it is perceived by another sense so that we cannot really say that it is either visible or tangible. The same applies to shape, size, number, movement which can be commonly said to be perceived but which in fact are perceived by various senses. Colour and light are however visible because we can perceive these by sight alone. These visible things because they are visible come within our eyes—I do not mean the things themselves but rather their forms—through a diaphanous medium as if it were through transparent glass. It is the water which exists in the pupil of the eye which renders the form visible and this happens because this water is enclosed in the eye, rather like a mirror which is glass backed with lead so that the form cannot travel any further but comes up against it like a ball and stops. Therefore the form which does not appear clearly in a transparent medium becomes bright when it is stopped for the same reason as an image appears when striking a glass backed with lead, and not when striking clear glass. From the pupil the visual spirit which continues from there to the fore part of the brain where is situated the sensitive power, immediately depicts the image and thus we see. In order that this picture should be true, that is to say similar to the original object itself, the medium through which the form travels to the eye must be without any colour and so must be the liquid of the pupil otherwise the visible form would be affected by the colour of the medium and of the liquid of the pupil. They who wish to make objects appear coloured in a mirror interpose the colour between the glass and the lead, so that the glass appears coloured. Plato and other philosophers maintained that we could see objects not because the visible reached the eye, but rather because vision left the eye to reach the object to be seen. This opinion however, is refuted by the Philosopher (Aristotle) in his book *On Sense and Sensation*. . . .

The above text perhaps is more understandable following our discussion in the preceding pages and gives us some idea of the way scholars of the time were thinking.

When we reach the fourteenth century we find a slow but continuous evolution in the sense that the metaphysical and philosophical discussions on the nature of *lux* tend to crystallize and therefore do not undergo any further noteworthy development. These new ideas soon became the classic rules which were introduced in classic texts and became the teachings of the 'great masters', and this explains why among the authors we find two eminent churchmen, namely St. Bonaventura and St. Thomas Aquinas. In the fourteenth century we find a new generation of thinkers who come nearer to a geometrical *perspectiva* as if at last Alhazen's influence was beginning to be felt.

One of the exponents of this new trend was Jean Buridan (*c.* 1300–1358) teacher at the University of Paris and its rector in 1327 and again in 1340. He exerted great influence on the philosophy of his time. He wrote many scientific works and in his *De anima* in particular he studied the problem of the knowledge of the external world with a progressive outlook. The main theme of Buridan's ideas is the negation of a hierarchy of knowledge to which many philosophers of the thirteenth century not only subscribed but gave great importance. For Buridan both the knowledge obtained through the senses and that obtained through the intellect were to be considered on the same level. This was of great value and of great significance for future developments inasmuch as it was the beginning of that movement which drew the interest of thinkers towards the mechanism of the senses and away from that of the psyche. From that time onwards this movement became gradually more important. With reference to the concepts discussed earlier on in this book, *lux* the 'psychic' light, was losing its value while the *lumen*, the physical light which until then had hardly been considered, gained in value. This was the line of thought followed by William of Ockham, born at Ockham (*c.* 1290–1349). He gave deep thought to the problem of vision in his work *In quattuor libros sententiarum*. The same subject was also discussed by Hervé de Nédéllac, or Natalis, in his *Quattuor Quodlibeta*. All these discussions were based on the concept of *species* but in each a slight difference was noticeable, indicating a very slow but continuous movement in the same direction namely towards stressing the importance of the physico-physiological process in the problem of knowledge in general and of vision in particular. In Nicholas Oresme's *De visione stellarum* the geometry of astronomical observations, with particular reference to atmospheric refraction, was developed as never before. The application of geometry to the study of optical phenomena is increasingly evident in the works of other scholars as for example the *Quaestiones perspectivae* written by Domenico da Chivasso and also in the book having the same title written by Henry of Langenstein (1325–1397) and particularly in the works of Biagio Pelacani of Parma. These philosophers are not very well known. Their works have been discovered in manuscript form, buried in various libraries, which seems to indicate that it had not been possible to print them either in their time or immediately afterwards. The ideas discussed in these manuscripts therefore must have been unpopular with the majority of their contemporaries. Once again within a very conservative and strictly controlled circle, an avant-garde movement was forming which was the embryo of what in a near future was to prevail and was to change radically the theories of vision and light. But this 'near future' was still two centuries away. The new ideas were not welcomed and so they received neither support nor encouragement. Indeed the tendency was to prevent the dissemination of these ideas so that they remained in manuscript form which, if not destroyed, lay forgotten in some library. Therefore it will be useful to give here some indication of the ideas which were prevailing in the fifteenth century.

A text of some interest is found in a sort of Encyclopaedia compiled by Reisch, prior of a Carthusian monastery near Freiburg. This work was widely known with the abbreviated title of *Margarita philosophica*, and there were at least fourteen editions between the years 1486 and 1583. The first had as its title *Epitome omnis philosophiae*, *alias margarita philosophica*, *tractans de omni genere scibili*. In it the nature of light is discussed but in a very peculiar way, especially since the names of Alhazen, Bacon and Witelo are not mentioned. Once again we find mention of images in the sense adopted by Lucretius, even though the concepts undergo important modifications. Thus the view is expressed that the image as a whole was not produced directly by the object. The image which was produced by the object in the medium immediately surrounding it, produced another image in a contiguous region and this in turn would produce another and so on until the eye was reached, unless of course, each image produced was fainter than the preceding one, in which case it would fade away. All this would happen in an interval of time which however imperceptibly small is not zero. These images would be endowed with a spiritual and immaterial essence and therefore they could cross and intersect each other in the transparent media without hindrance. A little further on we find rather different concepts:

> In the visible bodies we must distinguish between light (*lux*) itself, illuminating light (*lumen*) and colour. *Lux* is the natural property of luminous bodies that imparts a motion similar to that of the body to which it belongs. This movement is its essence and does not depend on anything else intrinsic in the body. *Lux* was given its existence by the Creator at the act of creation of the world when He said: '*Fiat lux et facta est lux.*' In fact before the creation of the 'luminaries' it fulfilled the function of the sun into which its nature passed on the fourth day.... *Lux* is in the sun inasmuch as it is luminous by itself; in other bodies it is only present by participation ... *Lumen*, namely the illuminating light, is the image of the light itself that is to say of *lux*, and its derivation is of a primary nature.... Colour is the extremity of the transparent body in the limited body, that is to say that colour is a quality that resides at the surface of a body that is both limited and opaque and which touches the transparent medium.... Among philosophers there is a widely held opinion that colour is not produced by light (*lux*) in its real and material essence, but that it exists in advance and is potentially visible when not illuminated and becomes visible only following the action of external light.... Philosophers do not agree as to the essence both real and material of colour. Some think that *lux* is the hypostasis of colour and of all that which is visible. They believe that colour is nothing more than the diversity of 'limitation' of light in diaphanous or opaque bodies as it can actually be seen from the colours of the rainbow, of the halo, of the neck feathers of peacocks or doves.... Other philosophers maintain that colour is born of a mixture of diaphanous with opaque and that the illumination itself is the action produced by the light (*lux*), first hypostasis of colours.... Some deny the real existence of the colours of the rainbow and of a halo ...

There are other opinions expressed which are not very different from these and on which there is no need to insist. What is more interesting to note is that the number of colours was considered to be limited to seven, namely white, yellow, orange, red, purple, green and black. All the other colours could be reduced to these. Another interesting passage is the following:

> the means of seeing is the transparent and permeable body which, at least in its last part where it is linked with colour, must be diaphanous in action because it is necessary that the colour should be excited by light so that it can then stimulate the sense of sight. In fact the colour, because of its materiality cannot enter into the sense of sight, it therefore generates a *species* of itself which is free from any material impurity and sends it to the eye by means of the *lumen*, namely in the image of the *lux*. . . .

Further on we find somewhat modified ideas:

> We call visual ray the *species* of the visible body. This is not a line nor a surface without depth, but rather a 'body pyramid' which has its base on the object seen and its apex in the seeing eye. A whole luminous body always 'outlines' a pyramid of its light at any point. This explains why in whatever point of space the eye may be, it can see the luminous object, but in different ways. . . . Some thinkers of good authority have said that vision takes place by means of an emission of a visual spirit which travels from the eye to the object seen. . . . There are others however, who think that the *species* or images of the objects seen and the spirit emitted by the eye meet in the interposed medium. . . . Since we try to adhere to the more commonly held views, we shall say that vision takes place by the reception in the eye of a *species* of visible objects by means of a pyramid which has its base on the object seen and its apex in the eye. As for the visual spirit that is lucid and clear it descends from the brain to the eye via the nerves. When it becomes modified by the images it returns with a confused sensation. On its return the soul awakes and in seeing the surface of its diaphanous material which is of the greatest purity and absolutely colourless, assuming a similar form to the object, it turns towards this object and sees it then distinctly. . . .

Here follows a similarity with the artist who fashions wax in order to obtain a final figure.

From the reading of this chapter of *Margarita philosophica* it is apparent that there was a general confusion of ideas that, even at the beginning of the sixteenth century, ruled the mechanism of vision and the question of the nature of light. Frankly we cannot discover the opinions of the author who, after having summarized the ideas of the various schools showed, in the end, that the old hypotheses of the Graeco-Roman schools were still holding their ground and that the most common opinion was in favour of the *eidola* or the *species* which were more or less corporeal and material and at the same time non-corporeal and immaterial. It is obvious that the great wealth of new ideas contained in the works of Alhazen and Witelo remained on the whole unassimilated and misunderstood. In this we must recognize the effect of the general conditions of the times and even more the effect of the dominating philosophical outlook. Alhazen had been an experimental physicist, even if many of the questions he studied do not belong nowadays to the realm of physics as we understand it. His outlook was that of a man continuously questioning Nature, observing and experimenting. In his work we find the description of instruments designed to study the reflection and refraction of light. Very few people could understand and follow his line of thought especially if we remember that in those days any experimental observation which contradicted Aristotle was not accepted by the learned men who considered Aristotle to be the authority. But, the Middle Ages were drawing to a close.

A sensational discovery which was to be the seed of profound changes took place towards the end of the thirteenth century namely the spectacle lenses. The actual date of this discovery has been narrowed down by modern studies to between 1280 and 1285.[6]

The history of these little transparent discs of glass is one of the most fascinating although it has been allowed to sink into oblivion because it was thought perhaps erroneously, that it did not bring any credit to science. However, when we study it thoroughly we discover its importance not only from the historical point of view. There is little doubt that the use of lenses for correcting presbyopia, as we call it today, was a discovery made purely by chance by

[6] See E. Rosen, *The invention of eyeglasses*, Journal of History of Medicine and Allied Sciences, XI, 1956

people who knew nothing about optics. It is enough to remember the ideas that existed when this discovery was made in the thirteenth century and which we have described in the preceding pages, concerning both the mechanism of vision and the structure of the eye. Let us add that nothing was known at the time about the ageing of the eye and about its loss of the power of accommodation by which the young and normal eye can see equally distinctly both near and distant objects although not simultaneously. Neither the mechanism of the visual rays emitted by the eye, nor that of the *species* presented any characteristics which lead one to suppose that an object very near to the eye could not be seen clearly. Therefore no one had the slightest idea of what alterations took place with age in the eye, nor what had to be corrected or modified in order to see near objects distinctly once again when the eye had lost the capacity of doing so unaided. On the other hand, in those days the knowledge concerning the law of refraction of the rays was still that prevailing in Ptolemy's time. In other words there was no clear knowledge of how the rays passed through a transparent plane surface and even less through a transparent spherical surface. Therefore no one could have the slightest idea of how a small glass disc, whose face or both faces might be part of a sphere, could act optically. There was no reason whatsoever to lead a student of optics to place in front of an aged eye a transparent little disc of glass whose surface was curved in such a way as to be thicker at the centre than at the edges. The only possible and reasonable explanation of the use of lenses for correcting presbyopia must be accepted to be purely due to chance. Perhaps someone making little glass blocks with curved surfaces for some purpose or other had noticed that although he was getting on in years, he could see near objects as clearly as he used to do when he was much younger. Very probably the person in question was an old master glass maker, engaged in fashioning those glass discs joined with lead which were used for windows at that time. It is natural that he would test them by looking through them at some definite structure, and in doing this he probably noted, not without some surprise, that his eye-sight was rejuvenated. Several considerations lead us to believe that this is a plausible reconstruction of the facts. In spite of diligent search it has never been possible to find any trace of who invented lenses for spectacles. This would seem to imply that it must have been someone who did not belong to that class of people who wrote about their discoveries. The actual locality of the invention is uncertain, the choice lies between the Arno valley and the Venetian lagoon, both being localities which at the time in question were renowned for their flourishing glass industry. The Italian word *lente* (lens) which was used at the time to indicate lentils, is a popular term and this by itself seems to strengthen the view that an object with such a name could only originate outside educated circles. If we want to be quite correct we ought to mention that the actual name given to these glass discs was *lente di vetro* (glass lentils) or 'crystal lentils' because if anyone had mentioned *lente* without qualifying the material of which they were made there would have been a risk of confusion with the vegetable. The best proof that lenses were not invented in a cultured environment is the way educated people treated them once their use was introduced. Lenses were thought unworthy of any attention and were ignored for over three centuries. No other conspiracy of silence was ever so unanimous and so lasting. In the whole period of time from the fourteenth to the sixteenth century we find very little mention of lenses, and the few who mentioned them were people well known for their lack of prejudice. No mention whatsoever is found in books and this indicates that lenses were not discussed in schools. Even the rare allusions that we find on the use of lenses in Alhazen, Bacon, Cardano and others, concern the use of lenses as a means of mangification but never as a means of correction of sight. Meanwhile, the use of lenses by the efforts of modest craftsmen, began to spread all over the world and in addition to 'full lenses', which nowadays we would call 'converging or positive', 'hollow lenses' or what we now call 'diverging or negative', were also introduced. The latter were applied to the correction of myopia although no one really knew what was the cause of this defect that was generally known as 'weak sight'. It is not known who was the first to make use of this important application, nor where and when. There must still have been unknown spectacle makers who were working purely by trial and error. There is very strong evidence that lenses for spectacles were not discovered by any theoretical studies or in the philosophical and scientific schools of the time. We would like here to anticipate the events by quoting from a book written in 1593 which is very little known in our days and yet is extremely important, because in it we find for the first time an attempt to explain the theory of lenses. The book in question is the *De Refractione* written by G. B. Della Porta (1535–1615). We shall have the opportunity later on of examining in more detail the truly valuable content of this work, but let us for the moment refer to a part in which the author attempts an explanation of how converging lenses can correct presbyopia.

> There are two reasons why older people by using convex lenses can see better and more clearly. First because with age the pupil becomes slack and not only the pupil but all the organs and the control of the organs of the body, which becomes incontinent. Because of the slackening of the pupil the rays wander more freely and carry to the crystalline lens the object less well defined. By means of converging lenses the rays of the simulacrum are once again re-united and the pyramid is more closely composed, as we have already discussed, so that converging lenses by constricting the simulacrum compensate the defect. The second reason is because in old people the vitreous humour becomes altered and less pure, as we have already proved in the first Proposition of this book, and when light enters the eye through a crystal it becomes clearer and brighter, and so the second defect of nature produced by catarrh is compensated.[7]

If we note that four centuries after the invention of lenses for spectacles this explanation was given by someone like Della Porta who was ahead of all others in the subject of refraction and who also had attempted to write a book on the theory of lenses, then we can be certain that the application of lenses to the correction of presbyopia was due purely to chance. Another reason in favour of the conclusion that the application of lenses to correct vision was purely fortuitous is supplied by the evolution of the studies in geometrical optics in the period between the thirteenth and the end of the six-

[7] See G. B. Della Porta, *De Refractione* Book VIII, Proposition XIV, Cantini, Naples 1593

teenth century. During this time the study of the phenomenon of refraction through a plane surface was continued, and indeed towards the end of the period mentioned we witness the beginning of the study of refraction through a curved surface also. Unfortunately the whole direction was wrong and the studies reached a blind alley. With the intention of studying refraction in what was thought to be the simplest case after that of a plane surface, scientists decided to study first the case of a complete sphere and then that of half a sphere. We know now that from the optical point of view the behaviour of a whole sphere is very complex because of the intervention of aberrations. There was no thought at the time of limiting the studies to only a small part of a sphere. When this step was taken, and we shall discuss this soon, optics made great progress. From all we have said so far about the study of optics at the end of the Middle Ages, it is obvious that in the thirteenth century no one had any clear idea of the way optical lenses worked. The general reaction of the mediaeval scientific world towards lenses is justified as follows. From the scanty literature existing about lenses when people began to write about them, we can detect the general diffidence of the philosophical world towards them. The word illusion appears almost continuously. The classic reasoning can be summarized thus. The aim of the organ of sight is to know the truth, namely the real structure of the external world, by representing to our mind the shape, the position and the colour of the bodies which constitute it. This takes place either by means of the visual rays, which emitted by the eyes go forth to explore the objects, or by means of *species* which are sent to the eyes of the observer by the objects when they are illuminated. The best way to learn the truth is not to alter either the rectilinear form of the rays or the regular travel of the *species*. The introduction of mirrors, prisms and lenses in their path brings inescapably an alteration of truth and these instruments make us see figures where the material objects are not and often make us see them enlarged or reduced, inverted, distorted, doubled and coloured. It is all a trick and an illusion. All optical means must be eliminated if we really want to reach the truth. No one considered lenses or mirrors, particularly curved mirrors, worthy of serious and conscientious study. This reasoning is the key which enables us to understand the attitude of the mediaeval scientific circles in the field of optics. Unfortunately nowadays it is completely overlooked by the student of the history of those times and many historical phenomena do not appear to have a rational explanation. The *Margarita philosophica*, which we have already quoted, and which is an encyclopaedia published at least three centuries later than the invention of lenses for spectacles, does not even mention them in spite of the fact that by then they were known all over the world and that scholars themselves probably made use of them for reading. The situation changes rapidly, however, as we approach the sixteenth century which was to prepare the way for great and radical changes in the field of optics.

Meanwhile Leonardo da Vinci (1452–1519), perhaps following in the footsteps of Alhazen and Witelo, built the camera obscura and what is even more interesting showed the analogy between this and the eye.

> The experiment which shows that objects send their *species* or images which meet the eye in the albugineous humour, can be proved when the *species* of objects illuminated penetrate through a small round hole cut into a very dark room. You will then receive the *species* upon white paper placed within this dark room and fairly close to the hole, and you will see all the aforesaid objects upon the white paper with their actual forms and colours but they will all appear smaller and upside down. . . . The same happens inside the pupil.[8]

This observation of Leonardo brings to a close the mediaeval period of the history of optics. During the thousand years discussed no new theories of light arose although we must admit that after the exhaustive work done by the ancient philosophers there was not much left to do. There has been some clarification of fundamental importance of the theories that survived. As we have already remarked, ancient optics had produced geometrical optics based on the postulate of the propagation of rectilinear rays, on the law of reflection and on experiments of refraction. During the

[8] Manuscript D of the Institute of France

Middle Ages this optics continued its development and in so doing it dissociated more and more the nature of the rays from any physical structure, and accentuated their hypothetical and geometrical character. In particular, the work on the behaviour of curved surfaces acting as reflective and as refracting surfaces was remarkable.

The contributions in the physiological field were of a fundamental nature because of the more accurate knowledge of the anatomy of the eye and, what interests us even more, because of the final demolition of the theory of rays emitted by the eye towards its exterior. By now only the mathematicians, followers of *perspectiva* and some nostalgic admirer of antiquity dared to remember that such a possibility had been discussed. The existence of a physical *lumen* external to the eye was already accepted unquestionably. It was a question of a 'something' emitted by the sun or by flames capable of illuminating bodies and making them emit the *species*. This 'something' could affect the human eyes to such an extent as to hurt them and could leave on them an impression which could even last for a short while and it was also capable of being reflected by concave mirrors and thus lighting a fire. Notwithstanding the long discussions which had taken place in the thirteenth century the ideas concerning the nature of *lux*, that is the light which can be seen, were less well defined. The nature of *lux* initially had reached a very exalted and almost divine position, but then gradually it began to lose value and interest. Its purely spiritual character was gradually weakened until it almost became something objective, a quality of the surface of luminous or illuminated bodies.

The mechanism of vision had not yet developed so far as to be considered as fully explained. In the studies of Alhazen however, there was the embryo of the concept which soon was to be accepted but which still was buried under the 'husks', the *species*, the simulacra and from which the western philosophers trained in the Graeco-Roman schools, could not free themselves. They attempted instead to reconcile these ideas with the revolutionary ideas of the Arab philosopher and they obtained a grotesque construction which was doomed to be changed soon after. The comparison between the camera obscura and the eye had already been made and this would soon lead to the desired solution.

Lenses for spectacles had definitely come to stay. Convex lenses for the correction of presbyopia and concave lenses for the correction of myopia. These small discs of transparent glass, prepared, made and distributed by very humble craftsmen, completely ignorant of the long scientific and philosophical discussions, began to multiply rapidly under the very nose of mathematicians who were sceptical and kept silent, and this prepared the revolution which was to change totally the attitude of philosophers towards optics. In all these studies the physiological charac ter tended to prevail but, while on one hand many observations were already beginning to give to physiological optics the standing of an autonomous science, on the other hand too much attention had been paid to 'optical illusions' and to *deceptiones visus*. To find that an instrument whose mechanism one cannot explain often gives erroneous or false indications necessarily leads to lack of confidence and doubt. This is so both for the naked eye and even more so for the eye aided by optical devices. On this question there is an interesting passage in the *Trattato della Pittura* by Leonardo da Vinci. The testimony of Leonardo is of great value because he was not a member of academic circles nor even a member of the higher culture, but simply a self-taught artisan albeit of great genius. His voice therefore is the voice of the man in the street when he states:

> The masters do not trust the judgement of the eye because it always deceives them for he who tries to divide a line by eye into two equal parts finds out that the experiment deceives him. Because of this the good judge has doubts but not so the ignorant.

This scepticism, this diffidence was the great flaw of mediaeval optics.

DE REFRACT. LIBER I. 15

I, & videbitur brachium B I ſumma peti in N, & paulo amplius deprimendo B, ſemper I eleuari videbitur , & tandem I ad O perueniet, ſic deprimendo , ſubleuandoque, voti demum compos fies , quia aſcendere & deſcendere imam partem videbis perpendiculariter .

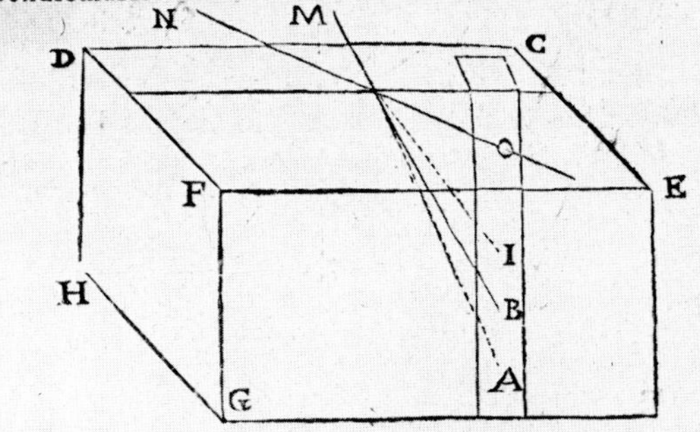

Satis tibimet ipſi verè facies , ſi aſtrolabium ex pelui extrahes , & regulam aliquam intra vas perpendiculariter erectam cõſtitues,& in imo repones obiectum valdè relucens,inde oculos deprimendo,videbis coloratum per virgulam aſcendentem perpendiculariter,ſic ſuſtollendo deprimi . Clarioris doctrinæ gratia exemplum apponã . Sit vas C D E F G, regula ad rectum conſtituta A B I O,viſus primo M,& punctus A videbitur in B, mox elongetur à vaſe oculus, deprimaturque,veniat in N, hũc punctum I in O videbitur loco ſublimiori, & ſenſibiliter aſcendere conſpicietur .

FIG. 14. The formation of images by refraction through a plane surface, from *De Refractione*

FIG. 15. Johann Kepler (1571–1630)

AD VITELLIONEM
PARALIPOMENA,
Quibus
ASTRONOMIÆ
PARS OPTICA
TRADITVR;
Potißimum
DE ARTIFICIOSA OBSERVATIO-
NE ET ÆSTIMATIONE DIAMETRORVM
deliquiorumq; Solis & Lunæ.
CVM EXEMPLIS INSIGNIVM ECLIPSIVM.
Habes hoc libro, Lector, inter alia multa noua,
Tractatum luculentum de modo visionis, & humorum oculi vsu, contra Opticos & Anatomicos,
AVTHORE
IOANNE KEPLERO, S. C. M[tis]
Mathematico.

FRANCOFVRTI,
Apud Claudium Marnium & Hæredes Ioannis Aubrii
Anno M. DCIV.
Cum Privilegio S. C. Maiestatis.

FIG. 16. Frontispiece of *Ad Vitellionem Paralipomena* by J. Kepler, 1604

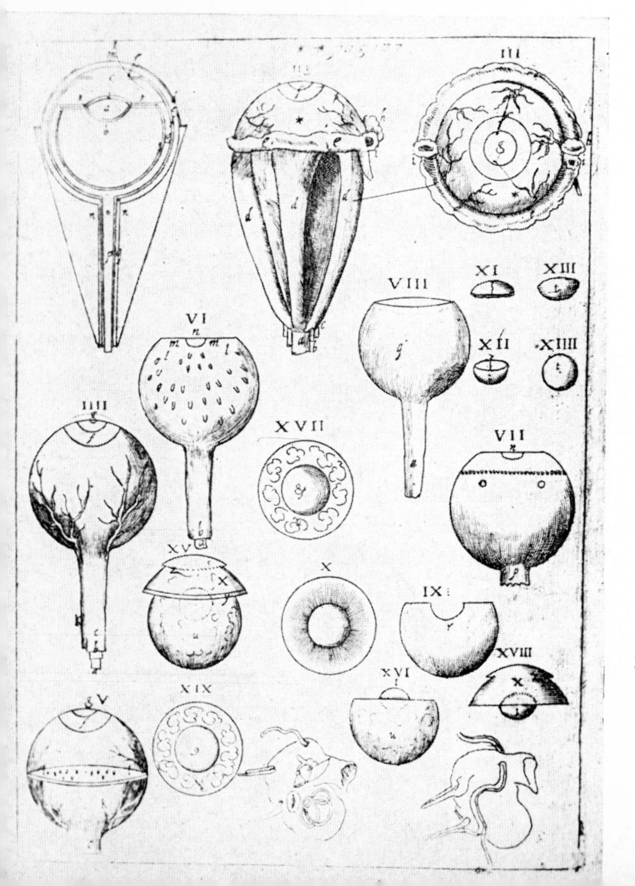

FIG. 17. Study of the eye, from *Paralipomena*

8 IOANNIS KEPLERI

bem omnem circumcirca illustret, quod fieri debere diximus. Sphæricum autem infinitas habet lineas.

PROPOSITIO III.

Lux seipsa in infinitum progredi apta est. Cum enim quantitatis & densitatis sit particeps, per superiora, nullâ amplitudine in nihilum abire poterit: quantitas enim, & sic densitas, diuisione in infinitum abit. Hæc de essentia. Sed & vis eiaculatoria infinita est, quia luci materia, pondus, seu resistentia nulla est per superiora. Infinita ergo virtutis ad pondus proportio.

PROPOSITIO IV.

Lineæ harum eiaculationum rectæ sunt, dicantur radij. Nam diximus affectari à luce figurationem Sphærici. Eius verò genesis verè Geometrica consistit in æqualitate interuallorum, per quæ punctum medium in superficiem diditur: Illæ verò sunt rectæ lineæ. Quod si curuis lux vteretur, nulla esset in didendo æqualitas, nihil igitur simile sphærico.

FIG. 18. The definition of luminous rays given by Kepler in *Paralipomena*

FIG. 19. Diagram illustrating the theory of converging central rays, from *Paralipomena*

FIG. 20. Diagram showing the pupil acting as a diaphragm, from *Paralipomena*

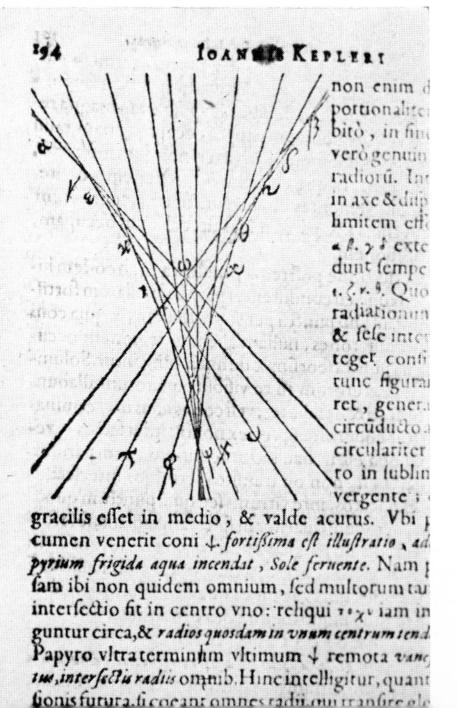
194 IOANNIS KEPLERI

non enim ... portionaliter ... bitò, in fine ... verò genuin... radiorũ. In... in axe & disp... limitem eff... α. β. γ δ exte... dunt sempe... ι. ζ. ν. θ. Quo... radiationum ... & sese inter... teget, confi... tunc figura... ret, genera... circũducto a... circulariter ... ro in sublim... vergente ...

gracilis esset in medio, & valde acutus. Vbi ... cumen venerit coni ♆. *fortissima est illustratio, ad...* *pyrium frigida aqua incendat, Sole feruente.* Nam ... sam ibi non quidem omnium, sed multorum ta... intersectio fit in centro vno: reliqui ... iam in... guntur circa, & *radios quosdam in vnum centrum tend...* Papyro vltra terminum vltimum ♆ remota *vane...* *tur, intersectis radiis* omnib. Hinc intelligitur, quant... sionis futura, si coeant omnes radii, qui transire gl...

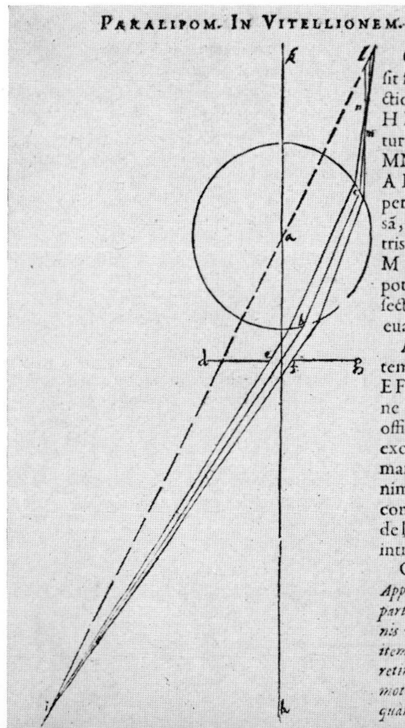
PARALIPOM. IN VITELLIONEM. 197

Cumq; in E F sit facta intersectio IE & IF. cum H K. terminentur verò radij in MN prius, quam A L transeant, per 19 præmissã, ergò ex dextris I fiũt sinistra M N. nec fieri potest, vt noua sectione dextra euadant.

Angustã autem fenestellam E F oportet esse, ne latior facta, officio hoc suo excidat: proximam globo, ne nimis parum & confusius quidẽ de hemisphærio intro radiet.

Corollarium.

Apparet hinc ex parte vsus foraminis vueæ in oculo: item, quare latera retinæ propius admota crystallino quam fundus.

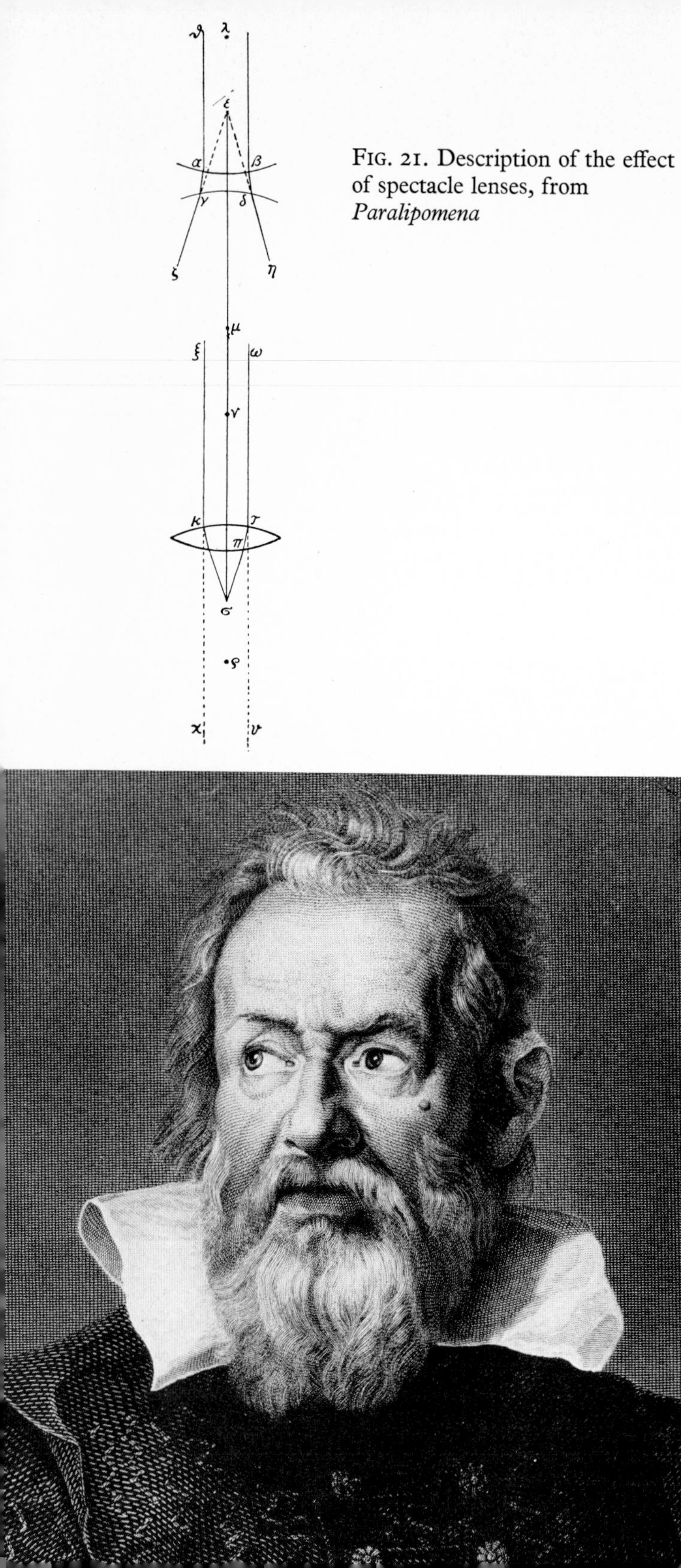

FIG. 21. Description of the effect of spectacle lenses, from *Paralipomena*

FIG. 22. Galileo Galile

CHAPTER THREE

The downfall of ancient optics

THE first half of the sixteenth century was uneventful in the history of optics. Mathematicians and philosophers studied the problem of refraction, both theoretically and experimentally; first the refraction through a plane surface and then through a spherical surface and in particular through a sphere of glass or a 'ball of glass' (*pila crystallina*). No conclusion of any value ensued. During this period the work of Maurolico da Messina (1494–1575) on optics and on vision had great influence but his writings were not published until much later and therefore we will discuss them later. The first event worth recording was the publication of *Magia Naturalis* by G. B. Della Porta.[1] This book had a wide circulation and enjoyed great popularity. It was translated into various languages and it had twenty-three editions in the original Latin, ten Italian translations, eight in French and others in Spanish, Dutch and even Arabic. The first edition which was written in 1558 when the author was twenty-three years old, consisted of four Books or Parts. The edition which interests us most is that published in 1589. In it the Books have increased to twenty. In Book XVII, which is devoted to optical 'magic', there is mention of 'glass lenses'.

Magia Naturalis is not a serious book. It is a collection of tricks, games and magic and this is the reason why Porta decided to include in it something about lenses. This however, had unexpected and important consequences which certainly the author himself had never imagined. Ronchi has studied in detail the contents of Book XVII of *Magia Naturalis* in another work[2] so that we shall limit ourselves here to the discussion of three points which have some bearing on our subject. The first is found in Chapter VI and consists of the description of the camera obscura with a lens at the opening and of the comparison made between this apparatus and the eye.

> Now I will reveal something that I have never disclosed and that I believed should not be told. If you fix a glass lens at the hole (of the camera

[1] See V. Ronchi, *Galileo e il suo cannocchiale*, Boringhieri, Torino 1964

[2] V. Ronchi, *Galileo e il suo cannocchiale*, Boringhieri, Torino 1964

obscura) you will see all things more clearly, the faces of men walking, the colours, clothes, actions and everything, and if you observe them carefully, you will have so much enjoyment that you will never tire of watching . . .[3] (Fig. 5.)

This experiment reveals to philosophers and to opticians where vision takes place, thus the question which has been discussed for so long, namely the introduction of images in the eye, is finally solved since it cannot be proved by any other way. The image enters the pupil in the same way as it goes through the hole of the camera obscura and the spherical lens in the middle of the eye acts as the screen, and this I know, will greatly please ingenious minds.[4] (Fig. 6.)

Cardano had also placed a 'glass lens' in front of the hole of the camera obscura. This we know from a brief passage in Book IV of his *De subtilitate*.[5] He described how he observed objects illuminated by the sun and which were outside a room, by means of an *orbem e vitro* inserted in a hole made in the closed wooden shutter of the window. Much more explicit however, is the description which the architect Daniele Barbaro gives us in his *La pratica della perspettiva*.[6]

[3] G. B. Della Porta, *Magia Naturalis* Salviani, Napoli 1589, Book XVII, Chap. VI

[4] Op. Cit.

[5] The oldest edition of *De subtilitate* is probably that published in 1547 at Bale but I have not been able to check whether it actually contains a mention of the camera obscura. I have, however, been able to verify this in the Paris edition of 1551 (ex officina Michaelis Fezandat et Robert Granion) and in the later ones. I must state that in every new edition there appears to have been some additions.

[6] Daniele or Daniello Barbaro was born in Venice in 1513 and died there in 1570. He was Patriarch of Aquileia. *La pratica della perspettiva* was published in Venice in 1568. On page 192 we find:

Natural means for showing perspective. Chapter V. 'Nature delights in teaching us the various proportions of objects and helps us to define the precepts of art, provided that we are diligent observers on every occasion. Now I will describe a most beautiful experiment concerning Perspective. If you wish to see how nature shows us the various

He tells us how a camera obscura with a converging lens in front of the hole is a means of obtaining accurate perspective when making panoramic and similar drawings.

Porta adds nothing to the subject which had not already been said by his predecessors. Had it not been for the world-wide distribution of his *Magia* only a very limited number of people would have known these ideas. As a result of this Porta's predecessors remained practically unknown and to Porta himself was attributed the invention of this device. As we have seen the comparison between the camera obscura and the eye had already been made by Leonardo. The relevant passage was in a manuscript which was only discovered many centuries later while the mention made by Porta attracted the interest of those who were capable of solving this question as we shall see before long.

Let us now pass to the second important point of Book XVII of *Magia*. The tenth chapter is devoted to 'lenses' as the title shows: '*De crystallinae lentis effectibus*'. For us the interest lies above all in

[6 contd.] aspects of things not only the outline of the whole but also of their parts as well as of their colours and shadows, you must make a hole of the size of a spectacle lens in the wooden shutter of a window of a room from where you wish to observe. Then take a lens from spectacles used by old men, that is to say a lens which is fairly thick at the centre and not concave like the spectacles for younger men who are short sighted, and fix this lens in the hole you made. After that close all the windows and doors of the room, so that no light is present except that which enters through the lens. Take a sheet of paper and hold it behind the lens and you will see on the sheet of paper every detail however small of everything outside the house and this will happen most distinctly at a given distance from the lens. By moving the sheet of paper towards or away from the lens you will find the most suitable position.' There follows a description of what can be seen and advice on how to obtain a good observation.

The above passage is extremely interesting not only because it describes the camera obscura with a convex lens in front of its hole with such details as are not to be found elsewhere, but also because here we find for the first time, at least as far as we know, mention of concave lenses used to correct myopia in young people.

the approach. After having specified that there were concave and convex lenses, Porta explicitly stated that the effects of the lenses were necessary for human life, and that so far no one had explained either the effects or the reasons:

> *... idem sunt et specillorum effectus qui maxime ad humanae vitae usum sunt necessarii, quorum adhuc nemo neque effectus, neque rationes attulit.*

Note that in the above passage the word '*specillum*' is used, which was the word adopted by scholars for lenses when they began to discuss or mention them. The use of this word confirms what we have remarked at the end of the previous chapter, namely that the word 'lens' was used by the general public and this would prove that spectacles originated in the workshops of craftsmen. Porta in his *Magia* almost always used the term '*lens crystallina*' because this work was not directed to scientists but rather to the general public. Only four years later he returned to the study of lenses in another work but this time more seriously and naturally here he only used the more dignified word '*specillum*'. The interest of the passage given above lies not only in the philological question of the word '*specillum*' but in the accusation which Porta levels against scientists for not having been able to explain how lenses worked; and not only for not having been able to explain the effects, but what is worse, for not having even examined or investigated them. It was the first time, after more than three centuries, that anyone dared to level such an accusation in a printed book. An accusation which, owing to the wide distribution of the book produced immediate effects.

The third passage which we shall quote from Chapter X of Book XVII of *Magia* is the famous text which, according to many, is supposed to contain the directions for the construction of the telescope with a concave eyepiece (fig. 7):

> ... concave lenses show very clearly distant objects; convex lenses show clearly near objects and therefore you can make use of them to help your sight. With the concave lens you can see distant objects as small but distinct, with the convex lens you can see near objects bigger but not well

defined; if you know how to combine properly these types of lenses you will be able to see distinctly and enlarged both near and distant objects . . .[7]

As we have mentioned, this passage taken by itself may suggest that Porta had put together a telescope of the Galilean type, as we would call it today. In reality examined in its context the passage refers only to the making of spectacles to correct some defect of vision which at the time was neither defined nor explained by opticians. But we shall soon see the consequences of this third passage. It is easy to understand that the discussion of lenses, of vision and of camera obscura in a book which became so widely known, broke finally the 'conspiracy of silence' which had been the only defence that scholars had against lenses. The accusation had been made and 'science' could not ignore it. Porta himself attempted an answer, this time as a scientist, in another book. This book, *De Refractione* already mentioned, is of great interest but has never been very well known even in his time. It was published in 1593, four years after the edition of *Magia* from which we have already quoted. The interest of this book lies in the fact that it is the first place in which an attempt is made to advance the theory of lenses. The author made a thorough study of the refraction of light (*lumen*) through a plane surface, through a sphere of glass (*pila crystallina*), through a half sphere of glass (which he called '*convexa sphaeralis superficies*') and through what remained of a slab of glass with parallel sides when a half sphere of glass had been removed, that is a plano-concave lens of very large aperture which he called '*concava sphaeralis superficies*'. Porta studied also the eye, the mechanism of vision, especially binocular vision, and showed the existence of many problems which neither he nor anyone else could solve, and finally in Book VIII he attempted to put forward a theory for concave and convex lenses. Book IX which dealt with colours and the rainbow, ends this precious little work. Apart from the historical interest of this first book on lenses, the whole work is important because Porta made use of the concepts of optics prevailing at his time and attempted to apply them to explain new

[7] *Magia Naturalis* Book XVII, Chapter X

devices and new experiments. In his attempt, which he represented as a success, he proved only one thing, namely that the classic concepts were absolutely insufficient and unsuitable to do this.

As we have already seen in the previous chapter, Alhazen had put forward the ingenious idea of considering an object as being made up of small elements each of which sent out its point-like *species* along a ray until it entered the pupil and reached the crystalline lens. By this mechanism he had overcome the great obstacle which had been such a stumbling block for earlier physicists who could not explain how the *species* of large objects could enter the very small pupil. Alhazen's idea however, was not fully appreciated by the western opticians who having made every conceivable attempt to reconcile it with classic ideas had reached absurd conclusions. Every object should emit its *species*, its simulacrum, in its entirety; perhaps to them it appeared absurd that the *species* of a man had to be divided into the *species* of his arms, his legs, and other parts as if in this dissection the *species* itself would lose its 'life'. The rays of the *lumen* would guide these *species* to the eye which was looking at the object. Such concepts could be accepted when it was a question of looking directly at another object with the naked eye. Complications arose when phenomena of reflection and refraction of the rays took place. So far, as we have already shown, the problem had been solved in a very simple manner when it was a question of plane and even curved mirrors, simply by not discussing it. Now however, Porta brought it to the fore again, involuntarily showing all the absurdities which followed the explanations of classic optics. Let us investigate this question in greater detail.

From the summary we have given it is easy to understand the significance of the first Definition with which Porta's work begins: 'The incident line is that along which the *lumen* of the source arrives or the simulacrum flows through a transparent substance.'[8] The author here considered two cases: first that of a luminous body, in essence the sun, which emitted its *fulgor* generally called *lumen*; and secondly, that of an illuminated body which emitted its own simulacrum. Both *fulgor* and simulacrum follow the incident line (note the singular) or what nowadays we call the incident ray.

This meant that the incident line was not an element of *lumen* or image, but represented only the direction along which travelled either the *fulgor* or the simulacrum, just as if it were a guide or a rail.

The second Definition stated: 'The refracted line is that along which travels the ray or the image in the second medium of different transparency'.[8] Ray here corresponds to *fulgor* of the first Definition, just as image corresponds to simulacrum. Here appears the word image which was to be the source of so many complications. Today we cannot understand how anyone could talk of refraction with such fundamental premises. Nevertheless it did not seem to worry Porta who imperturbably forged ahead confident of stating unequivocal ideas. It would be worthwhile to quote some of the arguments found in this book, but we shall have to limit ourselves only to mention some of the more outstanding examples. Let us remember that so far it was a question only of refraction through a plane surface.

Porta studied the phenomena which appear when a body immersed in water was observed obliquely from above. He concluded that the body appeared nearer to the surface of water and he explained this by means of the refraction of the ray, which reached the surface not along the normal to the surface but obliquely (we must remember that the observer was placed at one side). The refracted ray when it reached the air was bent by refraction towards the liquid surface (fig. 14). Today we would say that the refracted ray deviates from the normal to the surface of separation between water and air. Then the concept had to be expressed that the figure seen was in the direction of the refracted ray in the air produced forward in the water. Many knew that this was the case but the difficulty was to express this concept in terms of the optics of the time. Let us see how Porta attempted to explain this phenomenon. His second Proposition stated:

> An object when seen in a medium less perfect than air, if it is perpendicular, penetrates more strongly, if it is at an angle it deviates from the perpendicular.[8]

[8] G. B. Della Porta, *De Refractione* Cantini, Naples 1593 Book I

The Latin text is rather difficult to translate but we believe that we have given a faithful if not elegant interpretation of the author's idea. This is expressed in a particularly interesting way: the object (let it be noted) seen under water 'penetrates strongly' (that is to say without deviation) the air when it is perpendicular, but deviates from the perpendicular when it leaves the water obliquely. The expression, 'perpendicular object', indicates a manner of speaking which is not correct geometrically, but what the author had in mind was that: if the eye looks at an object immersed in water along the normal to the surface separating air from water, the object jumps out and enters the eye, but if the eye looks at it along a line of sight which forms an angle with the normal then the object jumps out but deviates from the normal. It is clear that Porta did not really mean that the object itself jumped out of the water, but it is not a question here of a simple mistake or of an abbreviated and conventional expression, the difficulty was to describe the phenomenon in another way as we can see also in some of the other propositions. For example, Proposition IV stated:

> The refracted image of the object running towards the eye is not seen in its place.

Here we have a change of ideas, it is the image and not the object which runs towards the eye and this is already a step forward. But there remained still many more steps to be taken. Where should this refracted image that travels (with the speed of light!) towards the eye be seen? Within the eye? Porta simply stated that it was not to be seen in its place. '*Imago rei refracta visui occurens, suo loco non videtur.*' According to Latin grammar *suo* should refer to the subject, namely to the image; supposing that it was an error of grammar, and it would not be the only one in the book, and that the author really intended to mean that the image was not seen at the place of the object, we are still left in a state of confusion. The following Proposition confirms the general confusion: 'The refracted image arrives at the eye along straight lines'.[8] A real puzzle follows. When the incident lines cross the refracting surface they are broken since this is the phenomenon of refraction. According to the first and second Definition given earlier on, simulacra follow both the incident and the refracted lines, but then the final simulacrum which reaches the eye, arrives following a straight line, namely a line which is not broken at the surface! Evidently the mechanism of vision accepted by the scientists of the time was faced with a test that was beyond it, and which strongly proved its insufficiency. Porta however, did not seem to notice this and carried on not the least perturbed. Naturally when he passed on to curved surfaces and to lenses the insufficiency appeared even stronger. There is no point in giving any further examples since we believe that what we have already discussed concerning the case of plane surfaces is sufficient.

In all the above discussion it appears clear that the union between image and ray as postulated is absurd. The *lumen* originated in luminous sources and fell on objects following rectilinear rays which could be reflected and refracted. It produced the detachment of images or simulacra from these objects and then guided them to the eyes of the observer. This basic concept is no more suitable for describing what happens in the case of refraction than it was in the case of reflection. The fact that it is useless except in the case of direct vision, when no complications intervene, leads to the conclusion that it is false and must be replaced. We can say that optics still lacked a light suitable to represent optical images. But events were close at hand which were to produce great changes.

While Porta was triumphantly writing what we can only call his absurdities concerning refraction and lenses, unknown spectacle makers had already produced, three years earlier, in 1590, the first telescope with a concave eyepiece. This happened in Italy the year after the publication of *Magia Naturalis* containing Book XVII on lenses and the alleged description of the telescope which we discussed earlier on. It is probable therefore that the telescope had been constructed just by a poor interpretation of the description given by Porta. All we know of this first telescope is that it appeared fourteen years later in Holland in the hands of local spectacle makers. We do not wish here to trace the history of the invention of the telescope, as this has been so amply discussed elsewhere, but soon we shall have to consider the profound repercussions that the instrument had on optics in general. Let us now examine the wonderful work of Johann Kepler (1571–1630), the great optician, mathematician and astronomer who laid the solid foundations of modern optics.

In 1604, that is only eleven years after the publication of *De Refractione* by Porta, there appeared the book *Ad Vitellionem Paralipomena* in which Kepler, under such a modest title exposed many fundamental concepts. The book consists of eleven chapters, the last six of which refer mainly to astronomical questions. The first five chapters are concerned with the nature of light, the basic principles of reflection, the localization of images, the measure of refraction and the mechanism of vision. All these subjects are treated in an orderly manner with principles which are completely different from those prevailing in the science of the time. Kepler, as the title of the work indicates, returned to Alhazen's ideas and developed them in their true essence without trying to reconcile them with the models of ancient optics or to adopt them. Thus free from concepts, which although two thousand years old now appeared insufficient and absurd, he could proceed rationally and logically towards a new structure of the intricate problem of light-image-vision. Let us see how he presents this wonderful construction. (Fig. 16.)

The first chapter deals with light: '*De natura lucis*'. The titles of the first four Propositions are:

> *Proposition I:* Light has the property of flowing or of being emitted by its source towards a distant place.
> *Proposition II:* From any point the flow of light takes place according to an infinite number of straight lines.
> *Proposition III:* Light itself is capable of advancing to the infinite.
> *Proposition IV:* The lines of these emissions are straight and are called rays.

Thus in these four Propositions is defined the luminous ray, which will later be finally adopted by geometrical optics. We should note here that contrary to the usage of the time, Kepler always used the term *lux* and not *lumen*. Later in passing he called attention to this distinction, but he did not follow it. This distinction had a deep relationship like that between cause and effect, indeed between a cause and an effect which are not necessarily and indissolubly tied, but linked through a psychological process that is still mysterious. At the beginning this uncertain link was a serious stumbling block, and making it a close link to the extent of identifying the cause with the effect, was a valuable simplification. Unfortunately this one agent which Kepler simply called *lux* continued to be indicated with one term only in the following centuries, and this did not contribute to a clarification of ideas. In modern languages we have words corresponding to *lux*, namely *light*, *luce*, *luz*, *lumière*, *licht* but no word exists to translate *lumen*, only in recent years an attempt has been made to introduce the usage of the word *radiation*.

The Propositions which follow in the first chapter (thirty-four in number) summarize the physicial properties of light and its relation with colour. Although at the beginning of the seventeenth century the knowledge of the principles of physics was more advanced than twenty centuries earlier, nevertheless it was not yet possible to co-ordinate them systematically into a plausible model. We can therefore repeat about Kepler what we have already said about Euclid, namely that while the geometrical content of his ideas represents a definitive and immortal conquest, the physico-physiological content suffers considerably from the scarcity of the experimental data and as a result it is incoherent and apt to break down. Kepler's *light* had infinite velocity.[9] As it went further from its source it was attenuated because it was spread over a larger area, but it was not attenuated in the direction itself of a single ray. A ray of light was pure motion and had no consistency, while light which moved was a surface. This strange idea, that light was a two-dimensional entity although it seems a prophetic intuition of what we now call wave-front, could not then be accepted. Proposition VIII is explicit on this subject. Under the title: 'The ray of light is not the light itself which travels', Kepler explained (fig. 18):

> In fact the ray, according to Proposition IV is nothing more than the motion of light. Just as in physical motion the motion is a straight line

[9] J. Kepler, *Paralipomena ad Vitellionem*, Frankfurt, 1604, page 9

but the physical entity which moves is a body, so in the case of light the motion is only a line and that which moves is a given surface.[9]

Kepler then proceeded to describe the transparency of *perspicui* or transparent bodies and that of *pellucidi* or diffusing bodies, the opacity of very dense bodies and then went on to define colour as 'colour is potential light, light buried in pellucid matter'.[10] In the Propositions which follow he explained that light was colourless and it only acquired colour when it was reflected by a coloured body. Light was reflected and refracted, that is to say that when the rays met a surface they were broken, but then acquired again their rectilinear form. Light could acquire colour also when it travelled through bodies and the thicker the body through which it travelled the more intense was the colour and if light travelled through media of various colours then it emerged with a new colour. If a body was illuminated by two lights of different intensity the weaker light could not be perceived. White bodies showed coloured lights which illuminated them, better than black bodies. Light was heat;[11] its heat was not material but it heated matter at least during a certain period of time. Therefore light destroyed and burnt bodies and in time discoloured them. These thermal actions were more intense upon black than white bodies.[12]

From what we have just seen the phenomena relating to light are more numerous and better defined than in any of the earlier texts on the subject. The geometrical concepts are also better defined and more advanced and had the latter been kept separate from the former they would have been a source of general admiration. Unfortunately the attempt to unite all the information into a single model cannot be said to have been successful. All this however, does not detract from the rest of the work because Kepler did not make use of any of the physical properties of his 'light' and he concentrated only on the laws of reflection and on the phenomenon of refraction, exactly as is done today in modern geometrical

[10] Op. Cit. p. 11

[11] Op. Cit. p. 25

[12] Op. Cit. p. 28

optics. Kepler concluded the first chapter of his work with a detailed discussion of the properties and nature of light as they had been propounded by Aristotle and as they were still being accepted by his numerous followers. In the following chapter the rays are used to explain the camera obscura which Kepler clearly declared to have learnt from Porta's *Magia Naturalis*. The rays are also used to explain those round bright spots due to the sun, which are seen under the foliage of trees and which had set a particularly difficult problem to the followers of the theory of *species*.

One of the most important of Kepler's theoretical conquests is found in the third chapter. 'Why is it,' asked Kepler, 'that when we look into a plane mirror do we see the shape of objects which are in front of it as if they were behind the mirror?' He began by summarizing the attempts made by his predecessors to explain this mysterious phenomenon and showed how unsatisfactory such explanations were. From this he went on to a more constructive approach under the title 'True demonstration' which he began with these words:

> In order to explain the real cause of the position of the images, the ignorance of which brings dishonour to a most beautiful science . . .[13]

He then proceeded with his explanation that we must first of all consider the images that we see as an 'intentional' entity namely subjective in the sense that is placed there by the eye of the observer. The eye receives the rays and from their structure it deduces where may be found the object which emitted them. Kepler first considered binocular vision and showed how the convergence of the lines of sight made it possible to determine the position of the radiant point noting that this in reality was a process of triangulation. He then considered single eye vision and called attention to the *triangulum distantiae mensorium* the triangle that had for two sides two rays emanating from a radiant point and directed to the extremities of a diameter of the pupil and this diameter formed the third side of the triangle. Kepler showed that this triangle supplied to the mind of the observer all that was re-

[13] Op. Cit. p. 59

quired to locate the observed point at the vertex of the triangle. It was clear that if the rays had maintained their rectilinear structure the image was seen at the place of the object; if, during their journey, the rays had been bent then the image was seen in a place which was not that of the object. This reasoning appears still clearer if we compare it with the confusion existing in Porta's *De Refractione*. It led to a conclusive and invaluable clarification of the theory of optical images obtained either by reflection or by refraction.

The whole of the fourth chapter is devoted to the measurement of refraction with long discussions on atmospheric refraction and on the history of this complex phenomenon. In spite of the great number of measurements Kepler did not succeed in finding a general law, but he only concluded that for angles smaller than 30° the angle of incidence and that of refraction could be considered to be proportional. This is the law that we still use today in elementary geometrical optics. The fifth chapter contains another of Kepler's important achievements. It is entitled '*De Modo Visionis*' and after an accurate description of the anatomy of the eye he stated:

> I say that vision takes place when the image of the whole hemisphere of the world in front of the eye and even a little more, is formed upon the concave reddish surface of the retina.[14]

For the first time after two thousand years of studies and discussions there was no hesitation in letting the light stimulus arrive directly to the retina. Kepler added that he left to the 'physicians' the task of explaining what happened after that, namely from the moment the image of objects reached the retina to the time when it was interpreted by the sense of sight, or by the 'soul' into the final act of vision. Today the scientists who are interested in this type of problem are no longer known as 'physicians' but as physiologists and psychologists but this is not important. What is interesting is the fact that Kepler at last overcame the prejudices concerning

[14] Op. Cit. p. 168

the inversion of the image on the back of the eye, namely the idea which had led his predecessors, following Alhazen's example, to consider the surface of the crystalline lens facing towards the pupil to be the sensitive surface. Kepler recognized the fact that the image projected on the retina was upside down but did not consider this inversion to be detrimental, for since the position of figures outside the eye was made by the eye itself, it could be associated by some rule or other with the distribution of the stimuli on the retina. The rule therefore was as follows: when the stimulus was low down on the back of the eye then the external figure had to be seen high up and *vice versa*; when the stimulus was on the right then the external figure had to be seen on the left and *vice versa*. This was a rule just as simple and plausible as that which said that we saw figures high up corresponding to stimuli of the upper part of the retina and to the right those corresponding to stimuli on the right side of the retina. Kepler was not content to suggest that the image of external objects reached the retina purely by overcoming the preconceived ideas about the inversion of the image, but he proved that this was the case. He examined the cone of rays having their apexes at the various points of an object and bases common to all in the pupil, in other words he examined the cones whose sections he had previously called *triangulum distantiae mensorium*. He followed the refraction of these rays through the cornea and the crystalline lens and showed that each one of these cones changed by refraction into another cone which had the same base in the pupil and the apex on the retina. This was indeed a very important step forward which finally left behind the optics of *species*. We have now a definite correspondence between the points of the object and those of the image on the retina, in place of all the vague and confused expressions which Porta had used so generously and which were in reality void of any meaning. To prove in detail the transformation of diverging cones into converging cones, Kepler studied the *pila* namely the ball or sphere, not the 'crystal sphere' but rather the 'aqueous sphere' and he studied its effect upon the cone of rays originating from a point of the object. He began by following the studies made by Porta to whom he made frequent reference and whose text he quoted. He repeated Porta's experi-

FIG. 23. Frontispiece of *Sidereus Nuncius* by Galileo; 1610

SIDEREVS
NVNCIVS
MAGNA, LONGEQVE ADMIRABILIA
Spectacula pandens, suspiciendaque proponens
vnicuique, præsertim verò
PHILOSOPHIS, atq; ASTRONOMIS, quæ à
GALILEO GALILEO
PATRITIO FLORENTINO
Patauini Gymnasij Publico Mathematico
PERSPICILLI
Nuper à se reperti beneficio sunt obseruata in LVNÆ FACIE, FIXIS IN-
NVMERIS, LACTEO CIRCVLO, STELLIS NEBVLOSIS,
Apprime verò in
QVATVOR PLANETIS
Circa IOVIS Stellam disparibus interuallis, atque periodis, celeri-
tate mirabili circumuolutis; quos, nemini in hanc vsque
diem cognitos, nouissimè Author depræ-
hendit primus; atque
MEDICEA SIDERA
NVNCVPANDOS DECREVIT.

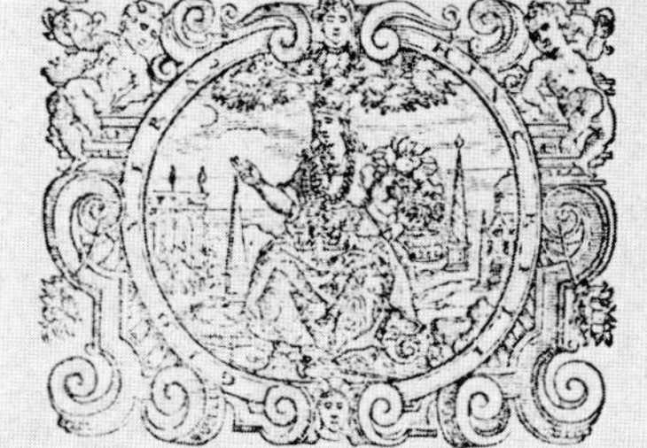

VENETIIS, Apud Thomam Baglionum. M DC X.
Superiorum Permissu, & Priuilegio.

FIG. 24. Passage describing the discovery of Jupiter's Satellites, from *Sidereus Nuncius*

celeris mihi pararetur instrumentum, (quod antea ob alterius
debilitatem minime contigerat) tres illi adstare stellulas
quidem exiguas verumtamen clarissimas; cognovi; quae licet e nu-
merum a me crederentur, nonnullam tamen intulerunt adm
eo quod secundum exactam lineam rectam, atque Eclypticae parall
ac ceteris magnitudine paribus splendidiores.
[illegible] videbantur, [illegible] illarum inter se et ad Iovem talis c
* * O * . ex parte scilicet orientali duae
stellae una vero occasum versus. Orienta
Occidentalis reliqua paulo maiores apparebant; de dis
inter ipsas et Iovem minime sollicitus fui; fixae en
[illegible] credite fuerunt. Cum autem die octava, nescio quo
tus, ad inspectionem eandem redirem, longe aliam constitutionem
erant a tres stellulas occidentales omnes a Iove, atque
viciniores, paribusque interstitijs mutuo dissepartae,

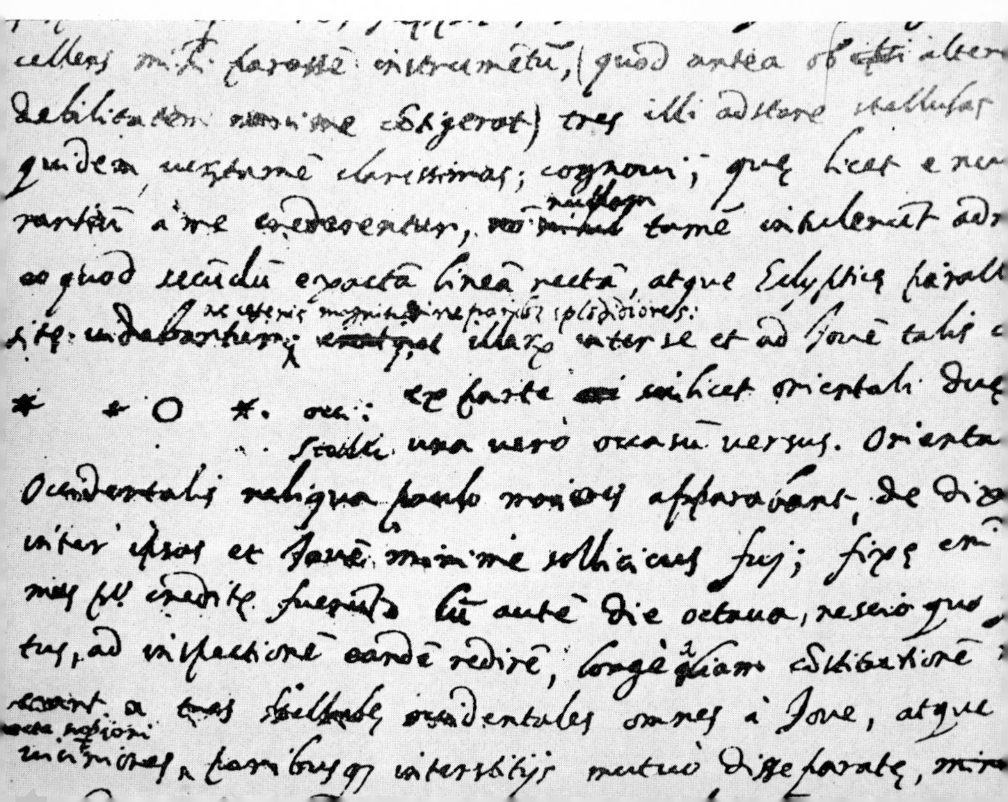

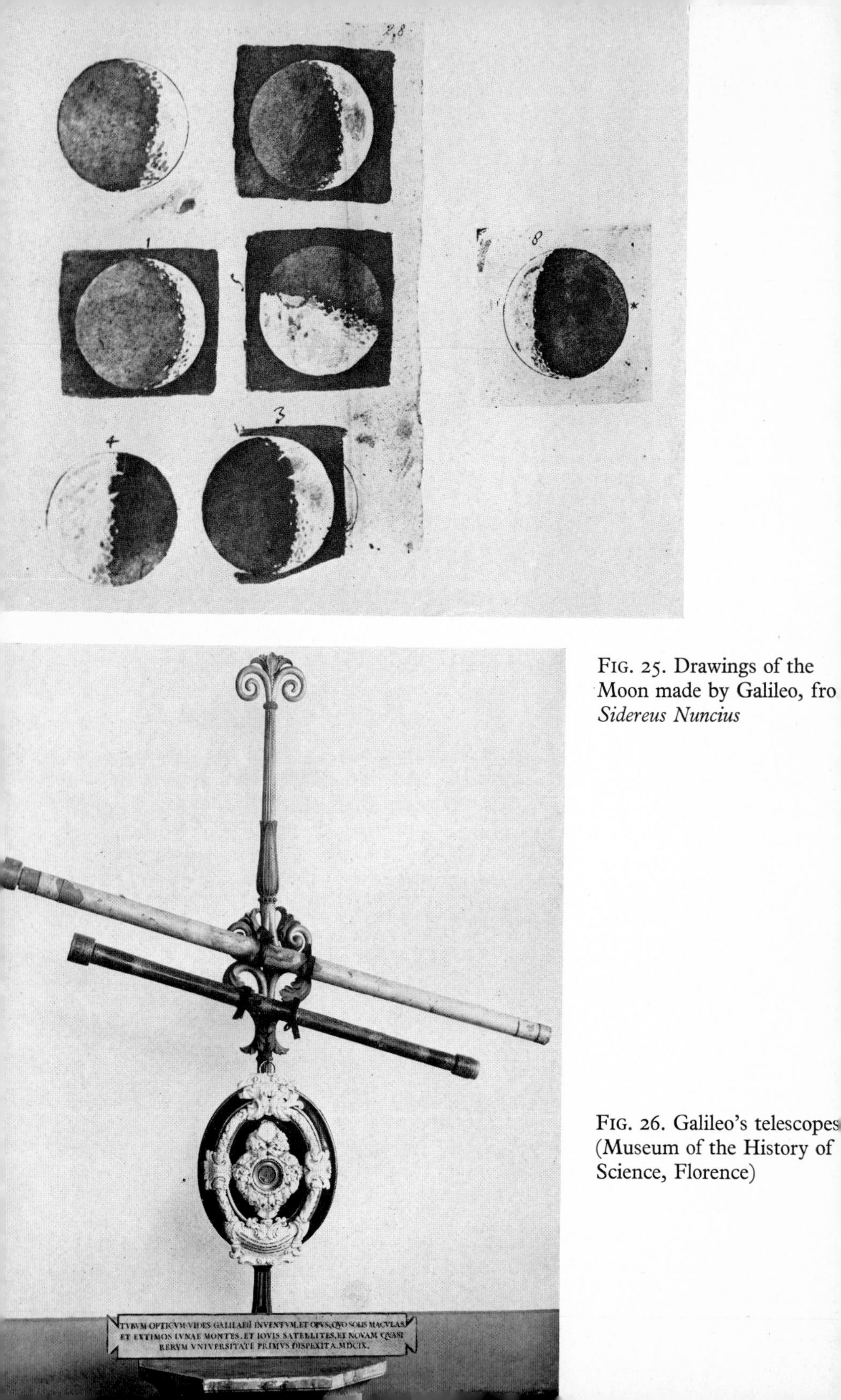

FIG. 25. Drawings of the Moon made by Galileo, fro *Sidereus Nuncius*

FIG. 26. Galileo's telescopes (Museum of the History of Science, Florence)

IOANNIS KEPLERI
S^r. C^æ. M^tis. MATHEMATICI

DIOPTRICE

SEV

Demonſtratio eorum quæ viſui & viſibilibus propter Conſpicilla non ita pridem inventa accidunt.

Præmiſſæ Epiſtolæ Galilæi de ijs, quæ poſt editionem Nuncij ſiderij ope Perſpicilli, nova & admiranda in cœlo deprehenſa ſunt.

Item

Examen præfationis Ioannis Penæ Galli in Optica Euclidis, de uſu Optices in philoſophia.

AVGVSTAE VINDELICORVM, typis Davidis Franci.

Cum priuilegio Cæſareo ad annos XV.

M. DCXI.

FIG. 27. Frontispiece of *Dioptrice*, by J. Kepler. 1611

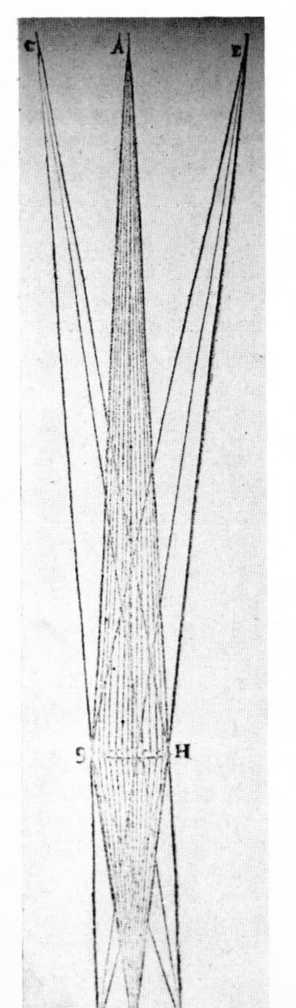

FIG. 28. Geometrical theory of lenses, from *Dioptrice*

XXCVI. PROBLEMA.

Duobus convexis majora & distincta præstare visibilia, sed everso situ.

Duo convexa sint sic disposita ad oculum, ut remotiùs solitariè ad oculum mittat imaginem eversam, non tamen distinctam, sed ut oculus lenti sit propior eo puncto in quo distincta repræsentantur, per LXXIIX. Vt si in schemate prop. LXXV. divergentia radiorum ab uno puncto D C, D P, ejusq; angulis O D P esset nimis magna pro oculo, oculusq; in O P esset extra D. F. puncta concursus. Interponatur deinde lens propinquior inter lentem illam priorem & oculum, hoc situ, ut oculus sit intra hujus punctum concursus, ut si in schemate Prop. L X X & LXXV oculus esset in I G. Quo pacto oculus per hanc lentem solitariam videbit erecta confusa itidem: sed ob causam contrariam, per Prop. LXXI. Ergò quia à remotiore lente, divergentia nimis est magna, hic jam à propiore convergentia contraria, illi nimiæ divergentiæ medebitur, ut ita corrigatur, & emendata accedat ad oculum ad distinctam visionem præstandam.

Et

FIG. 29. The passage describing the astronomical telescope, from *Dioptrice*

ABBATIS
FRANCISCI MAVROLYCI
MESSANENSIS.
PHOTISMI DE LVMINE,
& vmbra ad perspectiuam, & radiorum incidentiam facientes.

DIAPHANORVM PARTES,
seu Libri tres: in quorum primo de perspicuis corporibus. in secundo de Iride: in tertio de organi visualis structura, & conspiciliorum formis agitur.

PROBLEMATA AD PERSPECTIVAM, & Iridem pertinentia.

Omnia nunc primum in lucem edita.

NEAPOLI,
Ex Typographia Tarquinij Longi. M.DC.XI.
SVPERIORVM PERMISSV.

FIG. 30. Frontispiece of *Photismi de Lumine et Umbra* by Fr. Maurolicus of Messina, Ed. 1611

ments by observing with the eye through a sphere but soon realized that it was not the same thing to study *pendulae* images, that is images suspended in the air, or those which were collected on a screen. To distinguish between the two he introduced two different terms: *imagines rerum* were the images seen by the eye and *picturae* were those received on a screen.[15] The elimination of the eye of the observer with all its psychological complications was another innovation of great help in unravelling the tangled web of ideas concerning the formation of optical images. Kepler then defined the 'axis' of the sphere and studied the refraction of the various rays which cut the sphere at different distances from the axis, by using the approximate rule for refraction which we have already given. He noticed that of the whole beam of rays refracted by the sphere, those rays that were nearer the axis passed through the same point, while the peripheral rays did not converge so closely (fig. 19). Kepler then introduced a diaphragm in front of the sphere so as to act as the pupil and he observed that 'the image on the screen is inverted, but is very pure and extremely sharp at the centre' (fig. 20).[16]

The march towards the new optics proceeded from one conquest to another with unshakable logic. The *pila* with a very wide angle gave place to a sphere stopped down to a very small aperture and the beams of rays which were intermingled and in a complex tangle, were limited to their essential and ordered part so that they could be considered as small cones. At last, to a point of the object there corresponded a point of the image which was the apex of the cone of rays that had travelled through a sphere that had been stopped down to a very small aperture. This enabled Kepler to study near and far vision and to explain why people affected by presbyopia could see clearly near objects when using convex lenses which increased the convergence of the cones of rays and why people affected by myopia had to use concave lenses in order to see distant objects (fig. 21).

[15] Op. Cit. p. 193

[16] Op. Cit. p. 196

The sixth chapter of *Paralipomena* ends with the history of the studies concerning the mechanism of vision, and here Kepler gave Porta his due by recognizing him as his source of inspiration. He quoted in full the passage of Chapter VI of Book XVII of *Magia Naturalis* in which Porta affirmed that he had solved the question of how simulacra entered the eye through the pupil and Kepler declared that he started from these studies, eliminating the points which to him appeared erroneous. In this Kepler overestimated the work of Porta because his own contribution was not an addition or an improvement of the work of his predecessor, but it was a new, revolutionary approach, infinitely superior in every way. Compared with this, Porta's work was no more than a museum piece.

After 1604 optics changed its direction completely, but the immediate effect of Kepler's work was not very great. When we remember the great extent of the cultural environment, the antiquity of the views held by the scholars of the time, the tenacity and the faith with which many powerful Peripatetics defended the theories which originated with Aristotle, and when we remember also the great difficulties which surrounded the subject, we must not be surprised to find that several years after the publication of the *Paralipomena*, scholars still used the language used by Porta and by the earlier thinkers. They were still convinced that *species* and simulacra brought to the eyes the shapes and the colours of the external objects. Nevertheless, extraordinary new events tended to accelerate the evolution of ideas. The *Paralipomena* although a work of great importance had a serious omission, it made no mention of the theory of lenses. Only one page dealt, in passing, with lenses considering them as useful devices for increasing the convergence or the divergence of the cones of rays and therefore only suitable for correcting defects of vision. Therefore even for Kepler lenses did not have any particular interest as far as optics was concerned. It is interesting to note that he again declared that he was inspired by the work of Porta who was the only author so far, who had tried to explain how lenses worked.[17] Kepler stated that he was also indebted to Ludwig of Dietrichstein who for three years tried to interest him in the question of spectacle lenses.

Dietrichstein was a high official in Prague when Kepler was there. He was therefore a person of importance whom Kepler would not dare to ignore. The mention of him by Kepler indicates that the latter would not have undertaken such a study which at the time was considered by scholars to be a subject unworthy of study, had he not been compelled to do so to avoid offending a powerful local man. Very soon however, changes were to come. In the year the *Paralipomena* was published, Dutch craftsmen in Middelburg began to make telescopes which according to the recent investigations of De Waard, were copies of the original Italian model constructed in 1590.[18] The history of the invention of the telescope is outside our province but we are interested in the fact that when the telescope appeared it met with the general disapproval of the academic world for the same reasons that had been advanced by scholars against spectacle lenses. From a wide documentation collected in the book already mentioned[18] it is obvious that at the beginning of the seventeenth century opticians and mathematicians as well as philosophers knowing nothing of how ordinary lenses worked, could not be expected to understand the principle of the telescope. They summarized the value of this new device in the following manner:

> the telescope shows images larger than the real objects or nearer; it shows them coloured and distorted therefore it misleads us and cannot be relied to teach us the truth. Therefore it cannot be used as an instrument of observation.

The telescope, under these unfavourable conditions survived for several years. Spectacle makers tried to make telescopes and to sell them cheaply by using ordinary spectacle lenses and producing instruments of poor quality and construction. The situation changed radically five years later when a man who strongly influenced any scientific subject to which he turned his attention, became interested in the telescope. We are referring to Galileo Galilei (1565–1642). (Fig. 22.)

[17] Op. Cit. p. 200–1

[18] See V. Ronchi, Op. Cit.

Contrary to general belief, Galileo never spent much time in the study of optics. In a letter from Arcetri dated June 23rd, 1640, to his colleague Fortunio Liceti of the University of Bologna, he mentioned 'having always been in darkness' concerning the essence of light. More explicit is a passage from another letter written to the same colleague from Arcetri on August 25th of the same year:

> ... I am surprised that on account of what the philosopher Lagalla says, you should think that I considered *lumen* to be something material whereas you must have read in the writings of the same author that I have always considered myself unable to understand what *lumen* was, so much so that I would readily have agreed to spend the rest of my life in prison with only bread and water if only I could have been sure of reaching the understanding that seems so hopeless to me ...[19]

All this proved to be a great advantage to Galileo. When in 1609 he heard rumours about a device which showed distant objects enlarged and distinct, his mind was not clouded by the campaign of prejudice and of mistrust which the cultured society was waging against all optical devices and lenses in particular. Galileo undertook the construction of telescopes with his own hands and so was able to discover that lenses could be good or bad. When they were really good the new device opened new incredible fields to scientific research and to practical life. He started by modifying the reasoning of contemporary and past philosophers. It was true that the telescope showed images different from reality inasmuch it showed them larger or smaller, nearer or further, coloured and some time indistinct, but this did not necessarily mean that the telescope was always misleading because it was possible that from the images seen through the telescope the reality of things could be learnt better than with the naked eye. First of all for this result to be obtained it was necessary that the telescope should be perfect and in this case 'it can be of inestimable value'.[20] So Galileo was the first in the world of learning to reach the conclusion that we had to believe what we saw with the aid of the telescope. With this philosophical premise he turned the telescope to the sky and made

[19] The above letters are found in the *Edizione Nazionale delle Opere di Galileo* Vol. XVIII p. 208 and p. 233; 1937

wonderful discoveries which caused an upheaval in contemporary astronomy, physics and medicine. When he made his discoveries public in his wonderful little book *Sidereus Nuncius* he raised a storm of incredible violence. The whole academic world, with an impressive unanimity ranged itself against Galileo. He was accused of over-estimating and of accepting as true observations which being made only with a telescope, a mysterious and misleading device, were bound to be illusions. It was out of the question to claim to revolutionize science by means of such observations. (Figs. 24–26.)

Kepler who was asked to judge this question kept quiet for a long time, then he wrote a non-committal letter. He attempted to make some observations himself, but he was unable to build for himself a telescope which could show him the satellites of Jupiter and all the other astronomical discoveries made by Galileo. Only in August 1610, having made observations with a telescope constructed by Galileo and sent by him to the Elector of Cologne, did he realize the 'truth' of the new discoveries and he wrote to Galileo the famous words uttered by the dying Julian the Apostate: '*Vicisti, Galilaee!*'. From then onwards the victory of Galileo and of the telescope became total without any compromise. Naturally it took some years before the great majority of the academic world, with its inevitable inertia, could absorb the new ideas and change its attitude of mind. The change, when it came, was so radical that classical optics was destroyed and disappeared for good. Today a book on optics written earlier than the seventeenth century would be incomprehensible to the majority of people. So great was the upheaval caused by the triumph of the telescope that everything was new and changed. The history of this upheaval, which is so important and which is so little known today, has been given in the work already mentioned.[21] The student who wishes to obtain more details of the bitter struggle of sixteenth century science against the telescope, should read *Dianoia astronomica, ottica, fisica* by Francesco Sizi a Florentine knight. This is a very serious work

[20] *Edizione Nazionale delle Opere di Galileo* Vol. X p. 250, No. 228

[21] See V. Ronchi, Op. Cit.

in which are collected all the reasons why we should not believe what we see through a telescope. Sizi who was in a way the mouthpiece of all the scholars of the University of Pisa and of the Court of Florence, realized the seriousness of the situation when he said:

> . . . as buildings rest on foundations so science rests on principles. If these are undermined and destroyed, science like a building will collapse.[22]

He concluded his attack against the telescope by thanking God for having allowed him to reveal the truth. He had done much for the triumph of truth, but the truth which won was the very one he thought he had destroyed.

On August 30th, 1610, Kepler began his observations of the sky with the telescope which Galileo had sent to the Elector of Cologne. On September 11th of the same year he published *Narratio de observatis a se quatuor Jovis satellitibus erronibus, quos Galilaeus Mathematicus Florentinus jure inventionis Medicea Sidera nuncupavit.* (Description of the personal observations of the four wandering satellites of Jupiter which Galileo Galilei, Florentine mathematician by right of his discovery has called Medicean planets.) In this work he officially admitted that Galileo was right. In the same month Kepler presented to the Elector of Cologne for publication the manuscript of his *Dioptrice* which contained the theory of lenses on the same lines already developed in the *Paralipomena* on the subject of a sphere of glass or of water and of the eye. In this really precious little book, we find the geometrical theory of lenses such as we formulate even today, and a prediction of the astronomical telescope, and an explanation of the Galilean telescope and also the principle of the teleobjective. For four centuries no theory concerning lenses had been put forward because of the ban imposed by scientists and philosophers on lenses. From the moment Galileo infringed this ban only a few weeks were required for a theory explaining the behaviour of lenses. (Figs. 28 and 29.)

Before concluding the brief description of this important period,

[22] Sizi, *Dianoia astronomica, ottica, fisica* Ed. Naz. Opere di Galileo Vol. III p. 212

which had for protagonists three remarkable men such as Porta, Kepler and Galileo, we must mention an interesting case which deserves attention if we are to define the evolution of ideas in optics in the sixteenth century better and the value of those who contributed to it. It concerns the Abbot, Father Francesco Maurolico of Messina who died in Messina in 1575 at the age of 81. When he died Porta was 40 and Kepler was four years old. We have mentioned the ages of these three people because it is of interest to compare the work of Maurolico with that of the other two students of optics.

In 1611 a little book by Maurolico of only 84 pages was published posthumously, and consisted of two parts. The first has the title *Photismi de lumine et umbra ad perspectivam, et radiorum incidentiam facientes* and the second is entitled *Diaphanorum, seu trasparentium* and is divided into three Books or chapters. The book was published by Tarquinio Longo in Naples and in 1613 a second edition was published in Lyons. The cost of the publication was paid by a Genoese Maecenas named Airoli. The manuscript was reproduced together with some notes added by Father Clavius the well-known doyen of the Collegio Romano. The publisher tells us that Father Clavius' notes were added in italics. (Fig. 30.)

In the dedication to the Maecenas it is stated that it had been decided to publish this work at that particular time because the subject matter had acquired particular importance after the invention of the telescope. It was the wish of Maurolico's nephews to rescue from any possible accusations of plagiarism his writings which already existed in manuscript form often accompanied by errors and without Father Clavius' notes.[23] This little book is of particular interest because it is very different from the general books on optics of the time and has many peculiar similarities with Kepler's works. The plan of the book is as follows: after a few definitions and hypotheses, which we shall examine shortly, there is a discussion of the greater or smaller illumination which is obtained with rays of different inclination. Then the book studies shadows and the passage of rays through a hole, including an ex-

[23] See *Photismi* dedication. Published Longo, Naples, 1611

planation of the round spots produced by the sun under the foliage of trees. Finally there is a study of reflection by mirrors, plane and spherical, concave and convex, and even of cylindrical and pyramidal mirrors. In the three Books of *Diaphanorum* we find a discussion of refraction through a surface, through a slab of glass with parallel sides and a prism, a study of the *sphaera crystallina* and of the rainbow and, in addition there is a description of the anatomy of the eye, of the mechanism of vision and finally an explanation of how spectacle lenses work. In the last pages are collected twenty-four questions concerning optics and the rainbow (*Problemata ad persectivam et iridem spectantia*). (Figs. 36–38.)

The order followed is very similar to that of *Paralipomena* but when we examine certain details the similarity acquires a special character. For example in the hypotheses (*Supposita*) we find:

1. Every point of a luminous body radiates in a straight line.
2. The denser rays illuminate more intensely; rays of equal density illuminate equally . . .[24]

Clavius suggested that to these hypotheses should be added another namely that the rays should be slightly apart, as Euclid had already suggested. Instead Maurolico very definitely held opposite views because the first theorem which follows immediately these hypotheses states: 'Every point of a luminous body radiates every point of the illuminated object'.[25]

It is very remarkable how the concepts of Maurolico differ from those of Porta and how much closer they are to those of Kepler. It seems that Maurolico's concepts already contained the idea of luminous ray which we have seen so well defined in *Paralipomena*. Theorem XXII states: 'The forms of luminous objects seen through a hole are upside down on a plane'.[26] Evidently this is the camera obscura but it is remarkable that the author should make

[24] Op. Cit. p. 1

[25] Op. Cit. p. 2

[26] Op. Cit. p. 16

use of this reasoning to explain the round spots produced by the sun under the foliage of trees, just like Kepler. Theorem XXIII states that a surface illuminated by pure light emits a secondary light similar to its colour[27] just as in the *Paralipomena*. When dealing with concave mirrors Maurolico examined the reflection of rays originating from a point either on the axis or off axis. The geometrical studies previously made and which were described by Porta, agreed on one general incontrovertible conclusion, namely that the rays reflected at various distances from the vertex of the mirror cut the axis at different distances, so that they did not form a cone (the cone whose apex is called today point image) but they enveloped a curve called caustic (fig. 32) which geometrically speaking is an epicycloid. As we have already seen, Kepler had the ingenious idea of reducing the beam of rays by means of a diaphragm so as to use only the central group. Since these rays passed through the vertex of the caustic curve they formed a cone and its apex was the image of the point object produced by the mirror. Before Kepler, the reasoning that was considered to be correct was that the *species* of the object was reflected by the mirror and followed the path of one of the reflected rays as a guide. No one however, was able to say which one of the many rays. Porta's *De Refractione* is full of the absurdities which were the consequence of this reasoning. Maurolico's book does not mention the *species* at all. The object was considered to be point-like. From it many rays reached the mirror and all those which were reflected on a circumference which had its centre on the axis (what a modern student of optics would call a zone of the mirror) formed a cone which had its apex on the axis. Furthermore figures 33 and 34, taken from pages 27 and 28 of *Photismi* seem to express Kepler's new concept of images.

Finally, in the last theorem of *Photismi* Theorem XXXV under the title 'From the concentration of rays it is possible to produce fire', it is stated (fig. 35):

> In the concave mirror AB (fig. 29) from the point C of the solar body originate the rays CD, CE and CF which after reflection converge almost to a single point G. In fact although as we have already said

[27] Op. Cit. p. 19

> before, rays which reach the mirror from the same point after being reflected do not converge on one point, nevertheless the rays reflected by a small portion of the concave mirror meet almost at a same point . . .[28]

At the end of the theorem is given the date when this part of the work was concluded: 'Finished in Messina in the year of our Lord 1521 the nineteenth day of the month of October'.[28]

The reasoning expressed in the above theorem has nothing in common with that of Porta and in spite of the fact that it was advanced seventy years earlier it is much more rational and sensible. When we compare it with Kepler's reasoning we must admit that the latter has followed the identical path to reach his fundamental conclusions concerning the formation of images. Were it not that other circumstances led to a different conclusion it could almost be said that Kepler simply perfected and completed the reasoning of Maurolico.

The book *Photismi* ends with a page devoted to 'distorting mirrors' (*De erroribus speculorum*). The subject is not of interest for us but it is noteworthy that at the end of the page we find another date: 'June 13th, 1555, Feast of the Eucharist'. The author therefore, after thirty-four years had thought it opportune to make this addition to the *Photismi*. Let us now look at the three Books of the *Diaphanorum*. Among the hypotheses stated, the third and fourth are of interest:

> 3. If we multiply the angle of incidence, the angle of refraction is equally multiplied.
> 4. The object appears at the point where the visual ray produced meets the perpendicular line joining the object to the transparent plane.[29]

Here also Maurolico says more than Porta but less than Kepler. Porta limits himself to say that the angle of refraction is greater or smaller than that of incidence without saying how much. Kepler on the other hand, gives the same law of proportionality as given

[28] Op. Cit. p. 29

[29] Op. Cit. p. 31

by Maurolico but with the correct limitation, namely that the angles should not be greater than a given limit or, in other words, he calls attention to the fact that the law is not valid for large angles. The content of the fourth hypothesis is identical with the ideas expressed by Porta in *De Refractione*, but while the latter thinks that he has proved them and continues on with a reasoning that is without foundation, Maurolico presents the phenomenon as an experimental fact and he expresses it very correctly.

Continuing our examination of the theorems in Book I of the *Diaphanorum* we find a study of refraction through a slab of glass with parallel sides, through a prism and also through a biprism. In fact Theorem IV states: 'It happens that on account of the refraction of rays of an object, we see more than one'[30] and the author shows this with a drawing (fig. 36) which includes a biprism.

> Theorem VI: If the planes of the transparent body are not parallel then the object will never be seen in a straight line but according to rays which are broken on both sides at right angles.[31]

In the figure given and in the text the word 'right' (angles) has been changed into 'equal (*aequos*)' (angles). This is the first time that a prism is being studied in a book on optics. Next comes the study of the *sphaera diaphana*. The treatment here is very similar to that already used for the concave mirrors and at the end, Theorem XXIV, which is the last in Book I and which deals with the production of fire by the concentration of rays,[32] repeats that the rays refracted by a transparent sphere are concentrated almost at the same point (fig. 37).

Book II which follows deals with the rainbow. We do not wish to dwell on this subject as it is not directly connected with our study, but we must note that at the end of Book II yet another date is given: 'Evening of Sunday February 12th 1553'.[33] At that time no one had yet expressed the ideas about the rainbow which are

[30] Op. Cit. p. 32
[31] Op. Cit. p. 33
[32] Op. Cit. p. 48
[33] Op. Cit. p. 68

given in this work of Maurolico. He himself calls our attention to this in a note '*ad lectorem*' which ends Book II but he adds that it has just come to his notice that Canon Andrea Stiboni of Vienna had done similar work in this field. He leaves to the reader the task of comparing the two texts should he wish to establish the priority of the claim.

Book III is devoted to the study of vision. In it we find a description of the anatomy of the eye according to Vesalius, the famous Belgian anatomist, followed by an exposition of the mechanism of vision. Apart from the fact that in this part of the book we must take the word pupil to mean the crystalline lens, we notice a terminology which is very different from that generally used at that time. Maurolico still considered the crystalline lens to be the seat of vision[34] but he added something new which cannot be reconciled with the previous statement, namely that 'from its shape depends the quality of sight either short or long'.[35] He also added that the crystalline lens performed the function of receiving the species and of transmitting them to the optic nerve through the albugineous (vitreous) humour. The description of the anatomy of the eye is followed by a passage 'On lenses. How visual rays passing through a transparent body, convex on both sides, converge: and how passing through a transparent body concave on both sides diverge'.[36] Figure 38 explains this statement. The eight pages of this passage contain subjects of great importance. It is stated that the greater the curvature of the faces the more rapid is the converging or the diverging of the rays. After this the crystalline lens is studied as a lens. It is considered as the receiver of the *species* and at the same time also as the transmitter of them to the retina. For this purpose it is pointed out that its shape is not spherical but rather lenticular so that it does not invert the *species* but allow them to arrive concentrated, but the right way up, on the retina. The surface in front is intended to receive the *species* while the surface at the back is intended to transmit them.[37]

[34] Op. Cit. p. 69
[35] Op. Cit. p. 70
[36] Op. Cit. p. 73
[37] Op. Cit. p. 74

In studying this process, Maurolico examined the body point by point and the rays emitted by each point, in this way taking up again Alhazen's view that the perpendicular ray predominated over the others; nevertheless he remarked that these other rays also existed and entered the pupil. He concluded that before converging, and therefore before producing an inversion, these rays carried faithfully an image similar to the object 'through the cornea, the aqueous humour, the crystalline lens and the vitreous humour'.[38] Maurolico criticized his predecessors who went as far as letting the *species* reach the crystalline lens, but who did not then appear to concern themselves with the manner in which the *species* were to come out from the back surface in order to reach the retina. He seemed to find it natural therefore, that the various types of vision should depend upon the shape of the crystalline lens. A crystalline lens which was too curved produced short sight (myopia) while a crystalline lens which was flatter produced long sight. The shape of the crystalline lens and its variation was also responsible, according to Maurolico, for the difference in visual capacity between young and old people, namely presbyopia. Apart from explaining all this on the basis of greater or lesser convergence of rays he also gave an experimental proof by using lenses. Concave lenses corrected short sight and convex lenses corrected long sight because in fact they corrected the convergence of the rays therefore lenses of different curvature were required according to the degree of the fault to be corrected. Lenses of different curvature could also be used to enable people to read at various distances. He also stated that the age of the observer was linked with the various curvatures of the lenses which he needed in order to see objects that were near. After this Maurolico studied the lenses as means of concentrating rays and he extended explicitly to them the conclusions already reached for concave mirrors and transparent spheres. Although the rays when passing through a convex lens were not concentrated in a single point they were very nearly so. This could easily be seen by interposing a lens in a beam of sunrays which entered a darkened room through a narrow slit, the effect of convergence could be seen by the light diffused by illuminated

[38] Op. Cit. p. 76

dust. Just as in the case of concave mirrors it was possible to obtain a perfect convergence of the reflected rays into a point by substituting the parabolic shape for the spherical, perhaps it might be possible to make lenses of such a shape as to concentrate the transmitted rays exactly into a point thus obtaining a greater thermal effect.[39] A date is also given at the end of this chapter: 'In the fortress of Catania the 8th day of May 1554'.

We have made a rather detailed examination of this little book by Maurolico because it is really disconcerting. Apart from forecasting the possibility of making aspherical lenses, which Kepler did not even consider, the book contains many ideas which reappear later in the *Paralipomena*. We can almost say that Kepler has succeeded in completely unravelling a very tangled web whose threads had already been found by Maurolico. The fact that nobody, neither Porta nor Kepler, ever mentioned Maurolico's work and that Porta's ideas were less advanced, although in *De Refractione* he made superhuman efforts to be progressive, leads us to conclude that neither Porta nor Kepler knew the manuscripts in which Maurolico had expressed his ideas. On the other hand it is improbable that in the posthumous edition of Maurolico's work in 1611, the publisher had prepared the text by borrowing freely from the *Paralipomena* which had been published seven years earlier, because in this case surely the plagiarism would have been more complete. Moreover in *Diaphanorum* there are some notes which do not appear in the *Paralipomena*, for instance the mention of aspherical prisms and lenses. The text is such an organic whole and so logical that no new ideas, as new as Kepler's were compared with classic ones, could be inserted, because then it would have been impossible to maintain the order. Finally any such dishonesty must be discounted because of the evaluation of the work of Maurolico by Father Clavius, who was very well known and respected in the scientific world of the time. We must conclude therefore that Maurolico was an isolated forerunner who was not understood. What he wrote remained unknown because he was in advance of his time and no one even paid any attention to his ideas.

[39] Op. Cit. p. 80

Only thirty-six years after his death and after the triumph of the telescope and the downfall of ancient science, someone, perhaps Father Clavius himself, while re-reading Maurolico's manuscripts realized that Maurolico had understood many things long before other writers and this led to the publication of *Photismi* and of *Diaphanorum*.

This case is similar to that of Giovanni Rucellai, a Florentine nobleman, who in 1523 used a concave mirror to make a detailed study of a bee and described it in a short poem which was printed. This could have been done even two thousand years earlier but now that someone had attempted it and succeeded, we can ask why was there not an immediate development of microscopy by means of the application of concave mirrors and also with lenses? Rucellai remained isolated and was forgotten because he was not understood. He had made use of optical devices and, according to the current belief of the time, he was considered to have been a victim of illusions and deceit. Accepting that all this may be explained simply, what still remains a mystery is the extraordinary similarity between the reasoning of Maurolico and Kepler. It seems impossible that this should have happened without any link existing between them. Yet, as we have already mentioned, Kepler did not miss any opportunity to recognize Porta's achievements and indeed to state that his ideas developed from the views expressed in Book XVII of *Magia Naturalis*. Why would Kepler not have mentioned Maurolico as well if he had known his work? Why would he state that he had worked for three years at the request of Ludwig of Dietrichstein to explain the effect of spectacle lenses if all this had already been explained more than half a century earlier in *Diaphanorum*? There is no doubt that we must conclude that Kepler did not know Maurolico's work. The only possible link which could exist between the two may be explained by the fact that Maurolico was born in Messina, the son of a physician, a refugee from Constantinople, which had been invaded by the Turks. He therefore had some link with the East, indeed it is possible that his father knew Arabic and had read Alhazen. Kepler had studied Witelo and therefore indirectly Alhazen. It is possible that both Kepler and Maurolico had started from the same source and without being affected too much by western ideas they had understood and developed the important parts of Alhazen's work. Kepler who lived fifty years later, had some additional knowledge which enabled him to go further and indeed to reach definite conclusions.

The year 1610 marks the downfall of ancient optics. The *Paralipomena* when first published was not understood. What greatly changed the whole situation was the famous polemics about Jupiter's satellites. On one side was Galileo alone with his telescope and on the other the whole academic world. This was to become a struggle of life and death for scientific principles. The whole world awoke to watch and take sides in the struggle. Galileo's victory changed the attitude of mind of scholars concerning the eye and optical devices. At last a realization took shape that the excessive suspicion was a result of ignorance and that for centuries a powerful and valuable means of research had been unused.

With this change of mind, optics, which up to then had been left to only a few men devoted to scientific pursuits, became the centre of interest for philosophers and seekers after knowledge and became a topical subject. *Lux* now had a more definite interpretation than in the past. A *lumen* existed emitted by luminous sources and propagated in rectilinear rays which were emitted in all directions from each point of the sources themselves. This *lumen* was absorbed by opaque bodies, reflected by shining bodies, diffused by rough bodies and transmitted by transparent bodies and it carried heat with it. When it was pure it was white, but if contaminated by coloured bodies it extracted some colour from them. It was propagated with a velocity which, according to some, was finite and very great and according to others was infinite. This *lumen*, once it entered into the eye, converged on the retina and from there it stimulated the optic nerve giving rise to vision. As for the nature of this *lumen* some thought it to be pure motion, while others thought that it consisted of very minute particles which were moving with great velocity along rays. The question therefore had geometrical, physiological and physical aspects.

As we have already seen, geometrical optics was already estab-

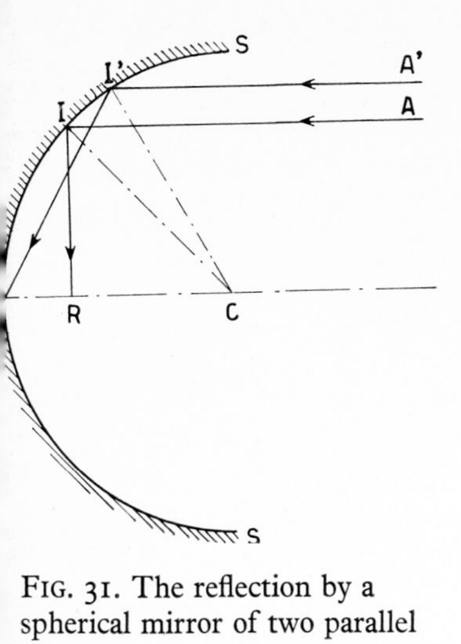

FIG. 31. The reflection by a spherical mirror of two parallel rays

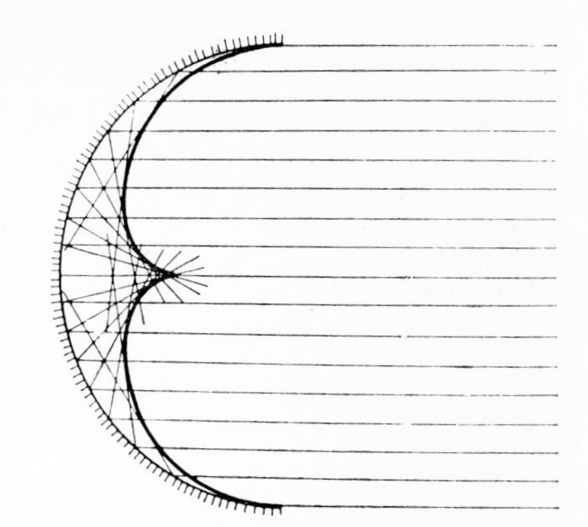

FIG. 32. Caustic curve

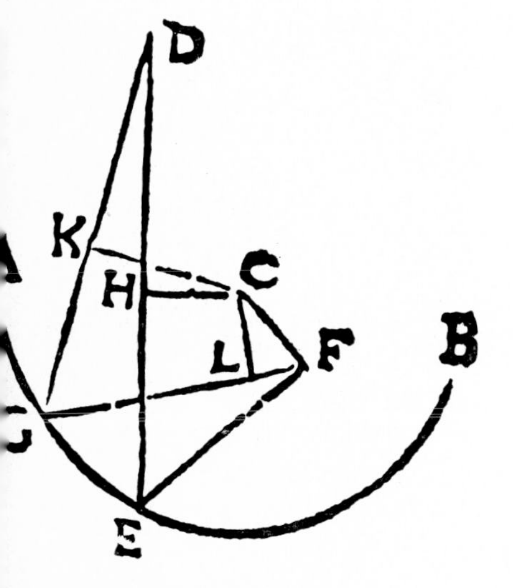

FIG. 33. Reflection from a spherical mirror, from *Photismi*

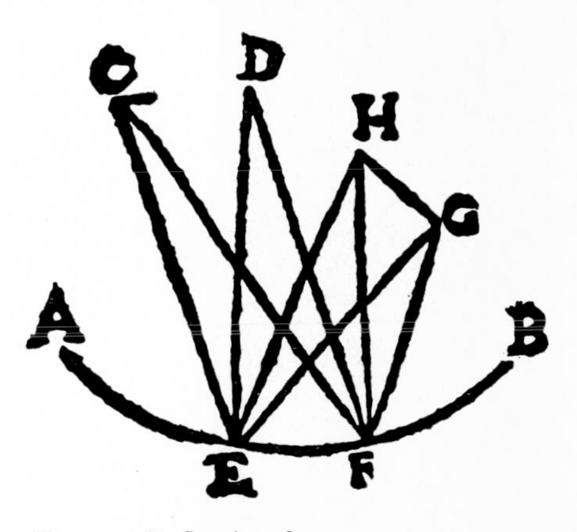

FIG. 34. Reflection from a spherical mirror of a double source, from *Photismi*

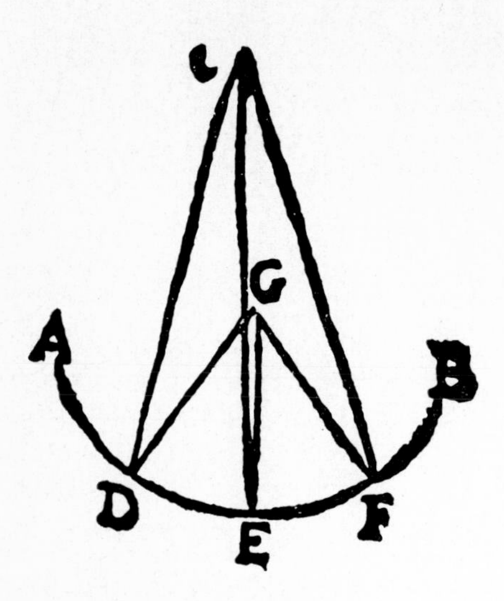

Fig. 35. Diagram showing that the rays reflected by a spherical mirror converge at a point, from *Photismi*

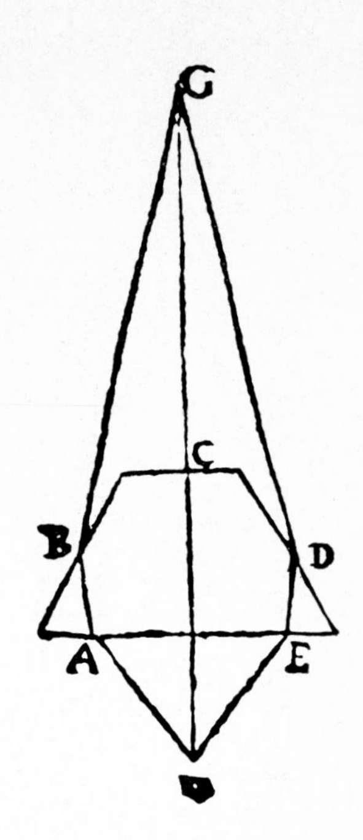

Fig. 36. The biprism, from *Diaphanorum*

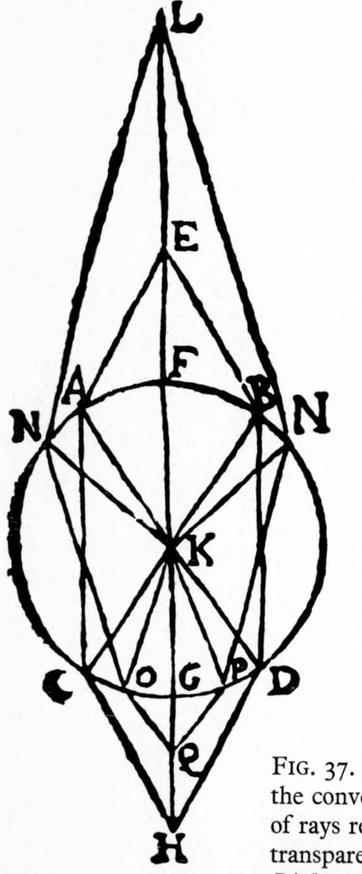

Fig. 37. Diagram showing the convergence at a point of rays refracted by a transparent sphere, from *Diphanorum*

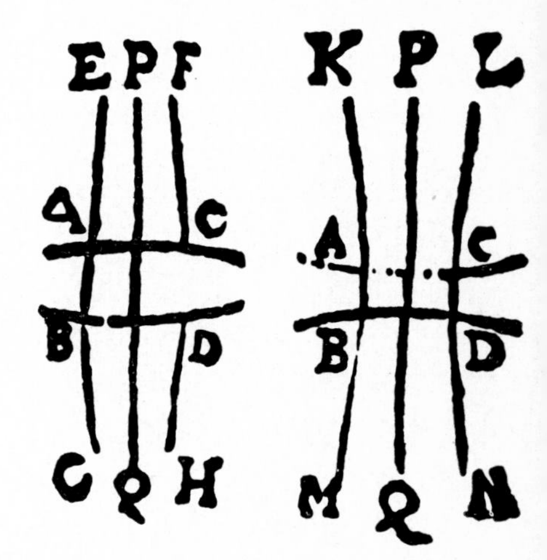

Fig. 38. Converging and diverging lenses, from *Diaphanorum*

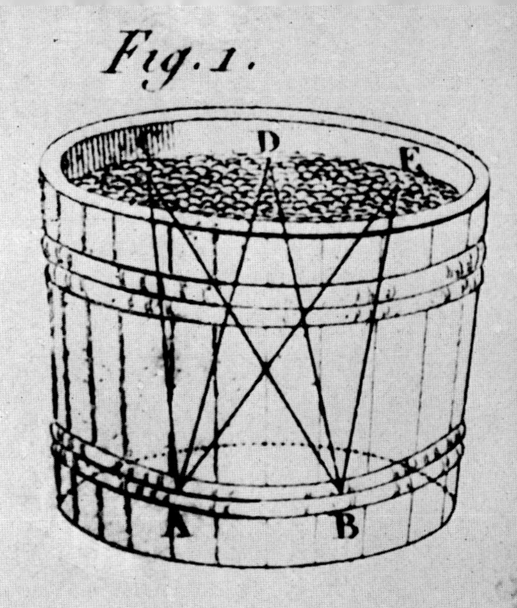

FIG. 39. Descartes' vat

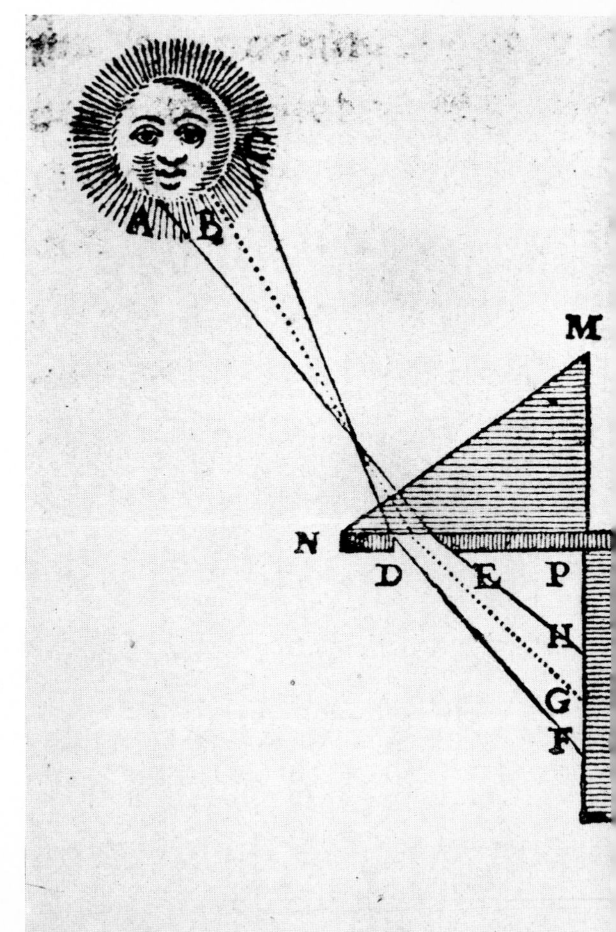

FIG. 40. Experiment with prism to show refraction and the formation of colours, from *Dioptrique* by Descartes

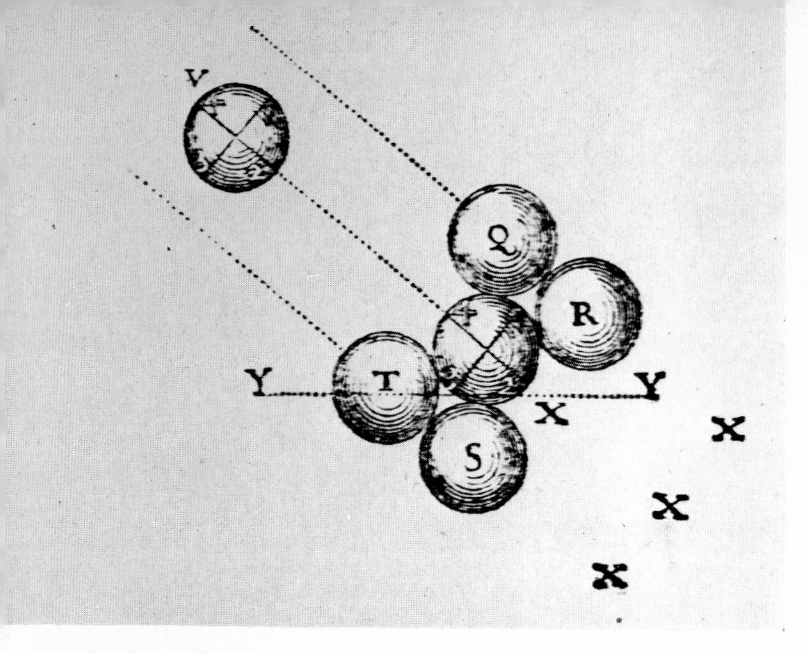

FIG. 41. Model illustrating the theory of colours, from *Dioptrique*

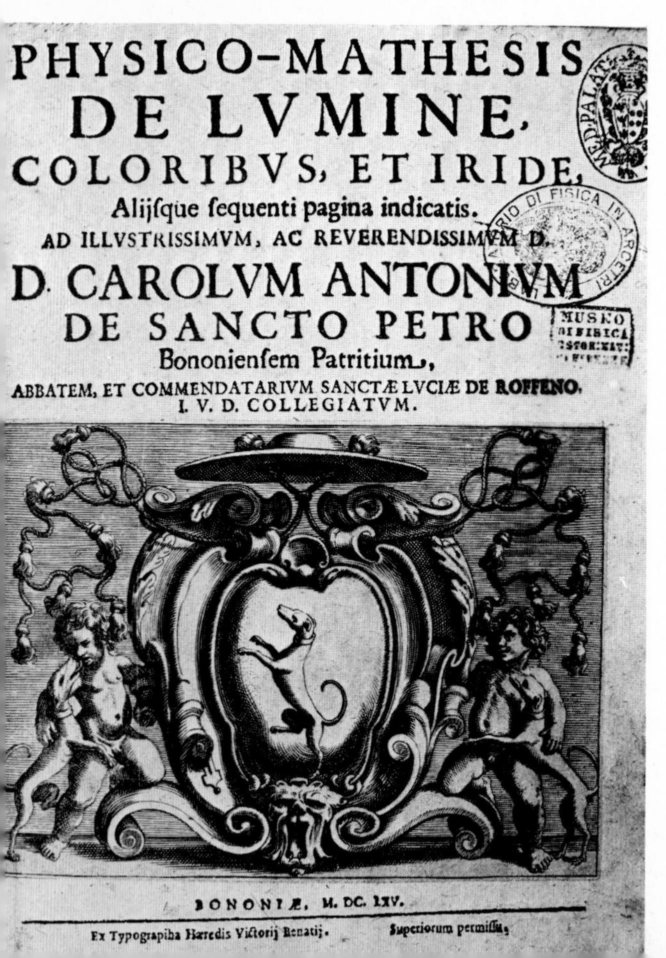

PHYSICO-MATHESIS
DE LVMINE,
COLORIBVS, ET IRIDE,
Alijſque ſequenti pagina indicatis.
AD ILLVSTRISSIMVM, AC REVERENDISSIMVM D.
D. CAROLVM ANTONIVM
DE SANCTO PETRO
Bononienſem Patritium,
ABBATEM, ET COMMENDATARIVM SANCTÆ LVCIÆ DE ROFFENO,
I. V. D. COLLEGIATVM.

BONONIÆ, M. DC. LXV.

Ex Typographia Hæredis Victorij Benatij. Superiorum permiſſu.

FIG. 42. Frontispiece of *De Lumine* by Fr. Grimaldi. Ed. 1665

lished in ancient times. Physiological optics developed in the middle of the Middle Ages and now to these two branches of optics there was added a third, physical optics, which was intended to study the nature and the properties of *lumen*. It is interesting to note that light was born because of vision and it could not be otherwise. Had it not been for the sense of vision we would not probably talk of light today. The ancient philosophers had approached the study of vision and of its stimuli with criteria which were similar to those adopted in the study of other senses. In the end the philosophers were divided into two schools of thought, which could be called objective and subjective and which reflected the eternal struggle between idealism and materialism. The clash between these two schools of thought concerned the eye and from the eye was to come the final proof in favour of one or the other of the two. This explains why it was so essential to solve the question of the mechanism of vision. Until a reasonable explanation was found of this wonderful phenomenon that enables us to see the shape of objects, we could not find the explanation of the other phenomenon just as wonderful which enables us to see brightness and darkness, nor could we decide who was right, those who believed that the 'husks' of objects entered the pupil, or those who maintained that the soul went out to touch the objects as if with a stick.

The seventeenth century began with a final clarification. The existence of *lumen*, which already in the Middle Ages had been considered as probable, was now accepted. The *species* and the 'husks' had disappeared and were replaced by beams of rays united in ordered cones. There still existed however, a great deal of confusion and perplexity concerning colour. Philosophers still persisted in their beliefs in optical illusions and in the deception of sight. The consolidation of natural philosophy and of experimental investigations silenced for ever these '*laudatores temporis acti*' and urged the physicists, full of faith and enthusiasm, to investigate the nature of *lumen* and to study its properties.

CHAPTER FOUR

From Descartes to Grimaldi

The seventeenth century began very brightly for optics. The question of the mechanism of vision was squarely faced, the telescope asserted itself, ancient optics was destroyed and the geometrical theory of lenses was well established. Events which were to follow were just as important.

In geometrical optics Snell (Willebrod Snell van Roijen of Leyden, 1591–1626) and Descartes (Réné du Perron, 1596–1650) succeeded in giving an exact form to the law of refraction and this became a very powerful weapon in the study of optical phenomena from the geometrical point of view and at the same time was an argument of great value in checking the various theories concerning the nature of light. As long as refraction was only partially understood, there was great freedom of interpretation and uncertainty about the nature of *lumen*, but now that the law of refraction had been stated, only the models of *lumen* which satisfied it could be accepted.

The question of the nature of *lumen* was the main purpose of the newly born physical optics. This new science was now acquiring great importance because, thanks to the definition of the mechanism of vision, the existence of *lumen* was finally accepted and also because the general trend of scientific studies due to the emergence of 'natural philosophy', gave great importance to physical rather than subjective principles. It was in fact on account of this philosophical movement that the interest in *lux* gradually diminished and indeed with time, it ended by being forgotten, while *lumen* became the only 'light' which was to be considered. At the beginning of the seventeenth century the same question was still asked as had been asked two thousand years earlier, namely what is light? Although the question had the same form, the meaning was very different. Once man used to say: There is light and there is darkness what are these two things that we 'see'? Now the question had changed. Man said: Outside there is a 'something' which is propagated in a straight line, which can be reflected, which can extract colours from bodies, which carries heat, forms images through lenses and stimulates the retina. This 'something' we call light, but what is it? This was the way that the seventeenth century faced the problem. All the philosophers of the time, no matter to which school they belonged, devoted themselves to this question. Today we tend to remember only Newton and Huygens and to consider them as the two great men who laid the foundation of physical optics. This is not really true and perhaps this tendency is due to the distance in time which as it increases tends to strengthen the contrast and to reduce the background. In reality, the discussion on the nature of light was fully developed even before these two men were born, and all the scientific world was busily engaged in trying to find an answer to this fundamental question.

If we wish to appreciate the importance of the contribution of Newton and Huygens we must first examine the contributions of other scholars. Unfortunately it has not been possible for us, however useful and desirable it might have been, to extend our investigations to the whole literature of the period in order to follow the discussions and the thoughts of every single philosopher, nevertheless we believe that we have reviewed the most important works.

We have already mentioned the name of Descartes. He had a well-deserved reputation and was respected by his contemporaries particularly for his work in optics. The study of the law of refraction, of some forms of aspherical lenses and of the rainbow, show the skill of the philosopher in the field of geometrical optics even though he had predecessors like Snell in the case of refraction and Antonio de Dominis in the case of the rainbow. It is time to say that Descartes did not study the question of the nature of light in depth. In his *Dioptrique* he discussed this question but he stated that for his purpose which was that of studying vision and telescopes:

> . . . there is no need for me to undertake to say really what is its nature and I believe that it will suffice to use two or three comparisons which help to understand it in the manner which I consider the most convenient to explain *all* those of its properties that experiments reveal to us, and to deduce from that *all* the others which cannot so easily be detected.[1]

[1] *La Dioptrique, Discours premier*. Oeuvres de Descartes, published by Victor Cousin, Paris, Levrault, 1824, Vol. V. p. 5

The first comparison is one which was made twenty centuries earlier, that of a man who in the dark armed with a stick can detect the objects surrounding him simply by feeling them with it. Descartes by analogy immediately deduced 'that light in the body we call luminous is simply a given motion or a given action which is very quick and lively and which moves towards our eyes passing through the air and other transparent bodies, in the same way as the movement or the resistence of bodies met by this blind man passes to his hand through the stick'.[2] This reasoning is not really new, but after this appears a concept which was very advanced for the times. Seeing colours was an operation similar to that by which a blind man with a stick felt trees, stones, water, etc., in other words colours were nothing more than different ways of moving or of resisting under the movement of the stick. Descartes began in this way his attack against the last form of '*species intentionales*', that is to say against the remanants of the 'husks' which philosophers continued to use in order to explain the nature of colours. As we have already mentioned, light, at the beginning of the seventeenth century was colourless. Light and colour were two different things. From the extracts we have quoted of *Margarita philosophica* in the preceding chapter and from what we shall see even better later in this chapter, philosophy had already ceded the study of the question of light to the physicists but had retained the study of colours, asserting that from coloured bodies, and in some cases even from white or black bodies, there flowed an emanation guided by luminous rays until it entered the eye to produce a stimulus. Descartes opposed this theory and with the arguments which we have already quoted he concluded that as nothing of the objects touched travelled along the stick of the blind man to reach his hands, the same must happen with light:

> . . . and so your mind will be freed from all those small images fluttering in the air and called *species intentionales* which clutter the imagination of philosophers.[3]

[2] Op. Cit. p. 6

[3] Op. Cit. p. 8

This statement may lead us to see Descartes as a follower of the Pythagorean school, but only a few lines after this definite statement we find him expressing some ideas which are completely Platonic:

> . . . objects which are seen can be sensed not only by the action which is in them and which comes to the eyes, but also by means of the action which is in the eye and goes towards the objects. Since however, this action is nothing more than light we must note that it is found in the eyes of those who can see in the darkness of night, such as cats; 'ordinary' men can only see by the action which comes from the objects . . .

These ideas are not very clear. Is light movement? Is it an action? Is light subject to modifications by illuminated objects and being thus modified does it enable the eye to see colours? What is there in the eye of cats and of those men who are not 'ordinary' which enables them to see in the dark? At this point Descartes realized that his first model or comparison was no longer sufficient and passed to a second; a vat full of grapes with two holes in the bottom. Since a vacuum could not exist in nature and yet porous bodies existed, these had to be filled by a very tenuous and very fluid matter, just like the must which filled the gaps between the grape pips in the vat, and these represented the coarser matter, which nevertheless did not prevent the must from flowing through the two holes in the bottom of the vat. (Fig. 39.)

> Thus all the parts of tenuous matter which are touched by the surface of the Sun facing us, travel in a straight line towards our eyes the very instant they are opened . . .

(according to Descartes the speed of light was infinite). The various currents did not interfere with each other neither were they disturbed by matter or by wind which agitated the air. It looks as if Descartes wished to give to light a material structure even by making use of a tenuous matter which filled the pores of real matter. Strictly speaking the real matter should now acquire a negative property acting as an obstacle or delaying agent although the example of the vat was used mainly to show that the must came out of the holes in the bottom even if there were grape pips in the vat. No one, included Descartes, doubted that if the fluid flowed through in spite of obstacles, it would flow better if no obstacles existed. We must confess that it is not clear how this concept agrees with that expressed earlier on, when Descartes compared the function of air and of transparent bodies with that of the stick or, in other words, when he considered them as transmitting the action that must reach the senses. Descartes did not seem to worry unduly about this and he used the second model to convey the idea of transparency and of the propagation of light in matter. He did not dwell on this and passed on to a third model that a projectile in motion and its trajectory were comparable to a luminous ray. With this he explained the reflection and diffusion on rough surfaces and he took as an example jets of sand formed of countless particles to represent a beam of light. Descartes did not stop to explain what was the relation between this third model and the other two. He seemed to forget temporarily the other two models and carried on with the third. Formerly colour was due to the manner in which bodies modified light, he continued with the same idea even if light, which until then was 'simply an action' now had become a projectile. The solution is extremely ingenious:

> We must note that a ball, apart from its simple and ordinary motion which will take it from one place to another, may have a second motion which may make it rotate around its centre, and that the ratio of the velocity of this motion to the velocity of the first motion can have several different values. When it is a case of reflection on shining surfaces the movement of rotation does not vary but in other circumstances it can change and then the colour of light changes also . . .

and he gave as example the experience of the man who plays tennis.[4] In discussing the various colours with reference to this model he added:

> . . . for I think I can determine the nature of each of these colours and demonstrate it experimentally, but this really is beyond the limit of my subject. . . .

[4] Op. Cit. p. 14

In his *Discours Second* of *Dioptrique* devoted to refraction, Descartes studied reflection, diffusion and refraction not of light but of projectiles. He deduced the law of refraction and of total reflection and he proved it experimentally thus:

> This has been sometimes demonstrated with unfortunate consequences when someone firing guns for fun into the bed of a river has wounded those who were on the other side of the bank.

Once he established the laws for projectiles, of course without taking gravity into account, he extended them to light with only slight variation:

> Finally as long as the action of light follows *in this* the same laws as the movement of this ball . . .

the conclusions reached for mechanics could be extended to light. This assertion is purely gratuitous and is only supported by a presumed experimental confirmation, for Descartes proved the law of refraction by following Alhazen's reasoning. He resolved the motion into two directions one normal and the other parallel to the refracting surface and he maintained the parallel component unchanged and varied the other on account of the resistance of the medium. Descartes added to it something new, a 'mathematical form' that the ratio between the sines of the angles of incidence and of refraction is constant. The fact that when light passed from air to water the ray was bent towards the normal to the surface of separation, led to the conclusion that the normal component of motion had increased! 'Perhaps you would be surprised if you carried out the experiment . . .' said Descartes himself. How did he explain the paradox, that the velocity of light should be greater in the denser medium?

> This however, you will cease to find strange if you remember the nature I attributed to light, when I said that it was nothing more than a given motion or action received in the tenuous matter which fills the pores of other bodies . . .

and in taking up again the comparison from ballistics he added:

> . . . as a ball loses more of its motion in colliding against a soft body rather than against a hard one, and as it rolls less freely over a carpet than over a bare table, so the action of this tenuous matter can be impeded much more by parts of air, which being soft and disconnected do not oppose much resistence, than by those of water which oppose greater resistance; and still more by those of water than by those of glass or of crystal . . .[5]

Once again we ask ourselves what did Descartes mean by light? Was it motion or matter, was it objective or subjective? We get the impression that he made use now of one model and now of another as a working hypothesis to represent rather than to explain, the properties of light. The conclusion at the end of his *Discours Second* is not easy to understand:

> . . . because finally I would say that the three models which I have just used are so suitable that all the peculiarities which may be noted in them correspond to some others which similarly are found in light; but I have only attempted to explain those which were related to my subject . . .[6]

Descartes is to be admired for his courage in saying this.

From an historic point of view we must call attention to an idea which is almost prophetic as it precedes by nearly three centuries a concept connected with the modern theory of relativity. In fact, still in the *Discours Second* we find the following passage:

> However other bodies may exist, mainly in the sky, where refraction produced by other causes have not the same reciprocity. Cases may well occur when the rays are bent even when travelling through a single transparent body just as the motion of a ball often is bent because it is diverted partly by its own weight and partly by the thrust received or by some other reasons. . . .

No one perhaps would think of Descartes in connection with the experiments to prove that light rays from the stars are bent when passing near the Sun on account of its mass. But to return to our subject, the ideas concerning light after the work of Descartes

[5] Op. Cit. p. 27

[6] Op. Cit. p. 29

were still confused and amorphous though showing some definite progress. The mathematical expression of the law of refraction was to be a great asset in helping the further development of geometrical optics and of experimental research, and in so doing helped the source of knowledge which was to define the concept of light.

The battle to liberate colour from the clutches of philosophers of the old school was engaged and the dilemma concerning the nature of light, material or kinetic, was beginning to make its appearance. These were only budding ideas, barely sketched it is true, but they indicated progress nevertheless. This is science.

The ideas of Descartes produced very interesting reactions. Among these perhaps the most important and worthy of note was the controversy which led Pierre Fermat (1601–1665) to state the principle still known as Fermat's Principle, which is one of the pillars upon which the theory of radiation rests.

The origin of Fermat's Principle, is indeed ancient. As we have already stated in Chapter I, the idea that 'sight must go as quickly as possible to the object to be seen' is to be found in the *Prospettiva* by Heliodorus who in turn had borrowed the idea from Hero of Alexandria. In fact he stated:

> ... as Hero showed in his book on mirrors, the lines which are bent at equal angles are shorter than all the other lines that originate in the same parts but are bent at unequal angles ... [He added] if nature has not been acting in a useless manner for our seeing, then seeing takes place by means of equal angles. This can clearly be seen since the rays of the Sun are bent at equal angles....

As we have already pointed out in Chapter I the concept of the shortest path had been proved to be correctly stated in the case of reflection, but it had perhaps been too hastily extended to refraction. At the time of Hero of Alexandria and of Heliodorus it was not possible to prove the validity of the concept of the shortest path in the case of refraction because at that time nothing was known about the speed of light. For this reason Hero's idea remained undeveloped until a very interesting advance in the seventeenth century. Pierre Fermat related how it happened in a long letter to an unnamed person, which appears in *Varia opera mathematica D. Petri De Fermat* (page 156), which was published in Toulouse in 1679 by his son Samuel, fourteen years after the death of the author. In this letter, Fermat relates how he came to give the proof of his Principle. The exposition is so clear and orderly that rather than telling the story in other words we feel that is better to give the text of the letter.

Sir,

Since M. de XXX speaks and you request it, you Sir whose reputation is so high and well established, I will awaken my Geometry which has been slumbering for such a long time and without more ado I would like to tell you the story of our *Dioptrique* and of our refraction, I will relate it as a tale thereby leaving your judgement free so that you may express your views without concern. After having seen the book of the late M. Descartes and after having examined closely the proposition which is the basis of his *Dioptrique* and which establishes the ratio of refraction, I became suspicious of his proof as his demonstration appeared to me to be a real paralogism. Firstly because he bases it upon a comparison and you know that Geometry does not give great weight to this method since comparisons in these cases are even more odious than they are normally; secondly because he supposes that the motion of light in air and in rarefied bodies is more difficult, or if you prefer, slower than motion in water and in other dense bodies, which to me seems to offend common sense; finally because he maintains that one of the directions or of determinations of the motion of a ball subsists entirely after having met the second medium, I also added some other reasons which would be superfluous or tiresome to relate. He saw my writings, he answered them and after a long correspondence and many arguments we parted like the accused and the witness, one asserting and the other denying facts, although I had in the end received very courteous letters from him.

After his death, M. de la Chambre having published his treatise on light and having honoured me by sending me a copy, I took the opportunity of writing him the letter which you have read, in which I suggested that the only way to avoid paralogisms in a subject so obscure, was to look for the explanation of refraction in this single principle, that is that nature always acts by the shortest path. On this basis I suggested that we could search by geometry for the point of refraction by reducing it to the problem or theorem which you know. Since however, I believed this method to be very difficult and very complex, because these questions of maxima and minima generally lead to long winded operations which can easily be confused by the unlimited asymmetry which is encountered in the work, I stopped my thoughts at this stage and for many years I awaited for some geometrician, less lazy than myself, to make either the discovery or the demonstration. But no one was willing to undertake this work. Meanwhile, every now and then I received letters from M. de la Chambre in which he urged me to add Geometry to my Principle and to put forward a demonstration which could serve as the real foundation to refraction. What was holding me back at first was the assurance given by M. Petit and others, that their repeated experiments to measure refraction in water, in crystal, in glass and in many other liquids, were perfectly in agreement with the proposition of M. Descartes, so that it seemed to me to be useless to try to find some other by means of my principle, since nature seemed to pronounce itself so decisively in his favour. The objection which you raise in your letter did not really worry me for the reason, which I had already stated in my letter to M. de la Chambre, that everything that touches or stops at a point of a curve is as if it touched or stopped at a point of a tangent to the curve at that particular point, so although the sum of two lines of reflection can be sometimes the greatest in mirrors that are concave, spherical or of any other forms, it is always the smallest of all those which can fall on the line or on the plane which touches the mirrors at the point of reflection and this does not require further proof since M. Descartes assumes it as much as I do. All the difficulty was therefore reduced to the fact that it appeared that I had to fight not only men but also nature. However, the recent insistence of M. de la Chambre was so pressing that two or three years ago I resolved to seek the help of my analysis believing that there were an infinity of propositions different from each other and whose diversity the sense could not verify so that I may happen to find one which could be very near to that of M. Descartes without being the same. I carried out my analysis by a particular method of my own and which Hérigone some time ago published in his *Cours mathématique*. I overcame all the asymmetry with difficulty, and at the end of my work, all of a sudden, everything became clear and I obtained a very simple equation which gave me the same proportion as M. Descartes. For a moment I thought I had made a mistake as I could not imagine that I could reach the same conclusion by a completely opposite way. M. Descartes had assumed for one of his demonstrations that the motion of light met greater resistance in air than in water, and I assumed exactly the opposite, as you can see in the copy of the demonstration which I tried to repeat from memory to give you full satisfaction, since the original had been sent to M. de la Chambre following my usual laziness. I repeated therefore, several times the calculations, by changing the data from which I started and I always came to the same conclusion. This confirmed two things; one that the opinion of M. Descartes on the law of refraction is correct and the other that his demonstration was faulty and full of paralogisms.

The followers of Descartes saw my demonstration given to them by M. de la Chambre and at first were inclined to refuse it notwithstanding the fact that I reminded them politely that they should be satisfied that the victory was still with M. Descartes since in effect his opinion was confirmed to be true even by reasoning different from his, after all the most famous conquerors did not consider themselves less happy when victory was obtained by auxiliary troops instead of by their own. At first they would not see reason, they insisted that my demonstration was false, because it could not stand without destroying that of M. Descartes which for them was above all others. When the more able geometricians who had seen my demonstration appeared to approve of it, then they congratulated me by means of a letter from M. Clairsellier, who is the gentleman who was instrumental in publishing the letters of M. Descartes. They called it a miracle that the same truth was reached at the end of two paths completely opposite to each other and determined to leave the whole thing undecided and to admit that they could not make up their mind whether to prefer M. Descartes or me, and that posterity would be the judge.

It is for you, Sir, who are destined by your exceptional merits to have great dealings with posterity, to inform it, if you think it right, of this famous contention or, if you prefer, to place this humble letter among your useless papers. I consent as I am indifferent. The same does not apply though to my very humble prayer that you should believe me yours etc.

To this letter is added a demonstration showing that the path followed by refraction, according to the known law of refraction, is that which requires the least time for the light to go from a point in the first medium to a given point in the second medium. Today any student of calculus could determine maxima and minima in an instant, but three centuries ago the ability of a Fermat was necessary to work out the problem by a long and tortuous way. This, however, is of interest to the history of mathematics rather than to the history of light.

Even if the involved mathematical reasoning was impeachable, the same could not be said of the physical elements on which it was based; and this is not said to point out the weaknesses of Fermat's reasoning but to show the basis of the developments which were to follow. The postulates from which the reasoning starts and which were used to set up the equations are expressed in an intuitive form which is unclear and ambiguous. We find mention of 'resistance' to the advance of light through various media and of time taken to cover certain paths, and there is an implied assumption that these two are linked by some simple law of proportion. Thus, while substantially we find reference to a tendency that nature has 'of acting by the shortest path', in certain parts of the work the demonstration concludes that the path travelled by light in order to go from a point B in air to a point H in water is that which takes the least time (fig. 44). This is demonstrated by postulating that the velocity of the propagation of light in air is greater than in water, and briefly that the ratio of the two velocities is equal to the ratio between the sines of the angles of incidence and of refraction.

It is interesting to note that at the end of the demonstration enclosed with the letter given above, Fermat did not hesitate to claim that the theory of Descartes had to be 'false' because the demonstration was valid only if the velocity of light was greater in air than in water. This shows us the fine diplomacy used by Fermat when he was suggesting to the followers of Descartes that they should be satisfied that the law of refraction was untouched by the polemics, and that they were still victorious. The followers of Descartes had realized that the victory was only apparent and not real, and as long as they could, they refused the very generous offer of Fermat. When they were forced to recognize it, they did so reluctantly and left the decision to posterity. A practice which appears to have been used later by Newton when necessary.

Fermat's letter, which we have given above, is without address or date, but it is fairly easy to determine approximately when it was written. The book of de la Chambre that we mentioned was published in Paris in 1662 by Jacques d'Allin. Fermat died in

1665, when he was sixty-four. In his letter he mentioned that he had carried out the demonstration about 'two or three years ago'. This was followed of course by a long argument with the followers of Descartes; so it seems likely that Fermat wrote the letter shortly before dying, perhaps at the end of 1664, or at the beginning of 1665. Probably the letter was never sent and must have been found after his death among his papers. This theory is strengthened by the fact that the demonstration was found together with the letter, which suggests that the letter was the original and not a rough copy since, by Fermat's admission, even when he had sent the first demonstration to de la Chambre he made no copy 'because of his usual laziness'. The important person whom Fermat addressed and who was charged with the task of informing posterity about the 'famous contention' (an indication perhaps that Fermat knew that the end was near) never received the letter and therefore could not carry out his wishes.

'*Substantia aut accidens?*'

After at least three centuries of discussion at the highest level, the problem which had been stated in this form by the *perspectivi* of the thirteenth century, came to the fore again. It is found in an extremely interesting and voluminous work entitled *Physicomathesis de Lumine, coloribus et iride* by Father Francesco Maria Grimaldi (1618–1663), published in Bologna in the year 1665. The work consists of a Proemio and of two Books. The introduction and the headings of the two Books immediately give a general idea of the content of the work. (Fig. 42.)

> Book One—Consisting of sixty propositions from which, by means of new experiments it is possible to deduce what seems to be favourable to the opinion of some concerning *de luminis substantialitate*. This however, will be refuted in Book Two and on this occasion many questions will be dealt with relative to apparent and permanent colour and many others related to the rainbow will also be discussed.
>
> Book Two—Consisting of six propositions in which it is shown how and for what reasons the Peripatetic opinion about the *de luminis accidentalitate* can be held. However, from all this it does not follow

FIG. 43. Francesco Maria Grimaldi (1618–1663)

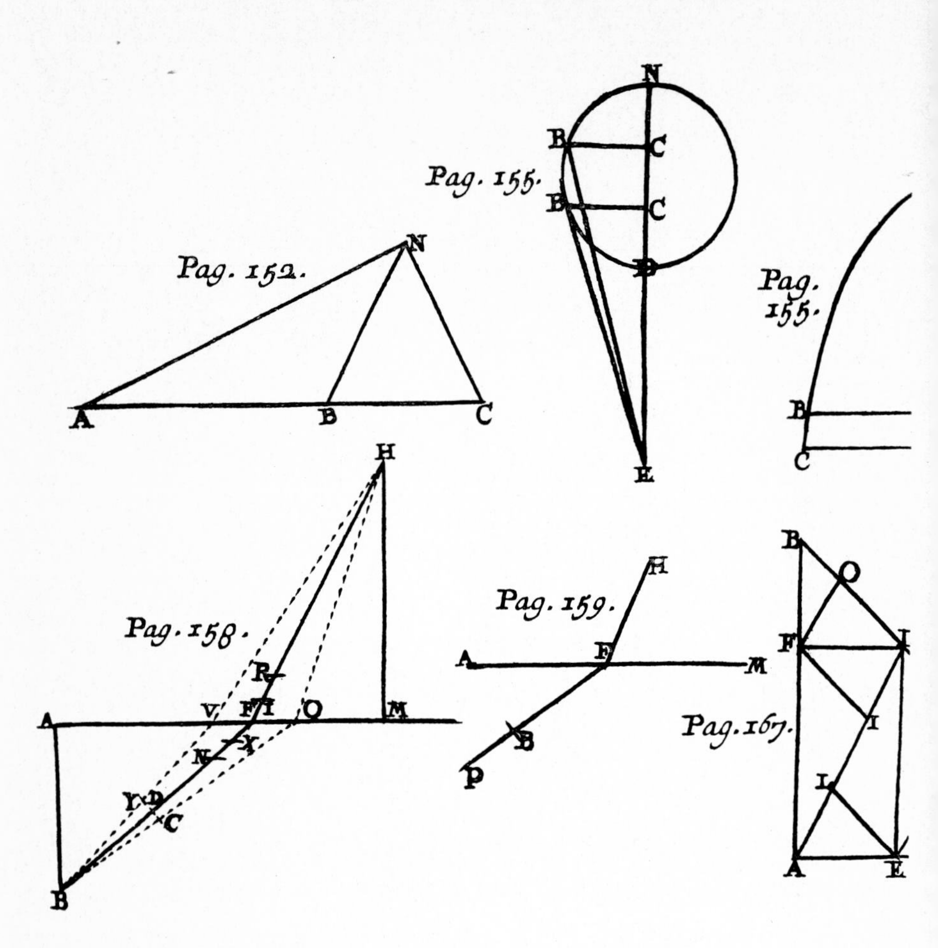

FIG. 44. Fermat's Principle and the law of refraction

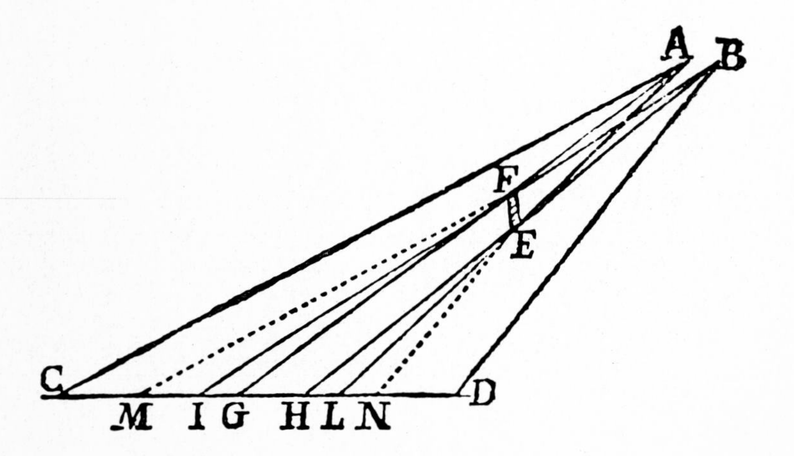

FIG. 45. The first experiment on diffraction, from *De Lumine*

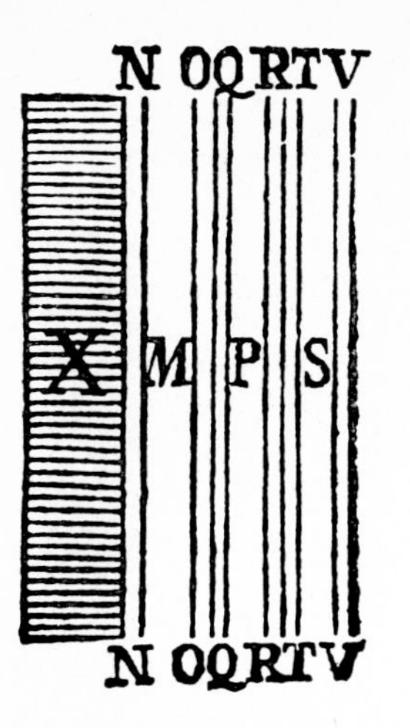

Fig. 46. Diffraction fringes at the edge of a shadow cast by a rectilinear object, from *De Lumine*

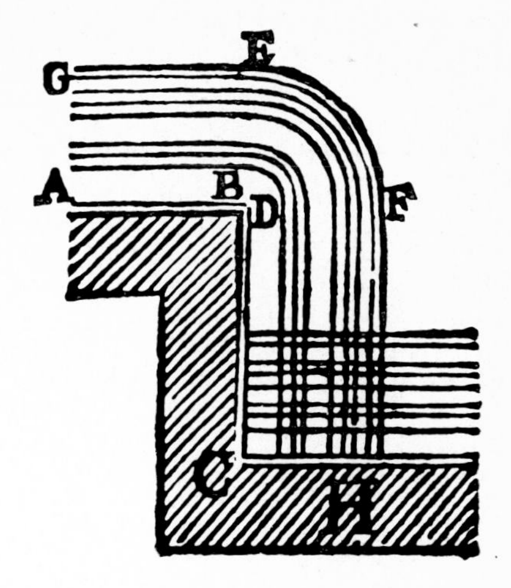

Fig. 47. Diffraction fringes at the edge of a shadow cast by an irregular object, from *De Lumine*

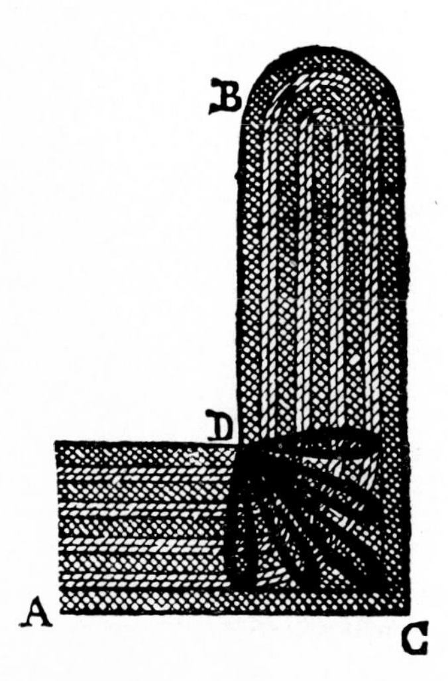

Fig. 48. Diffraction fringes in the shadow of an object, from *De Lumine*

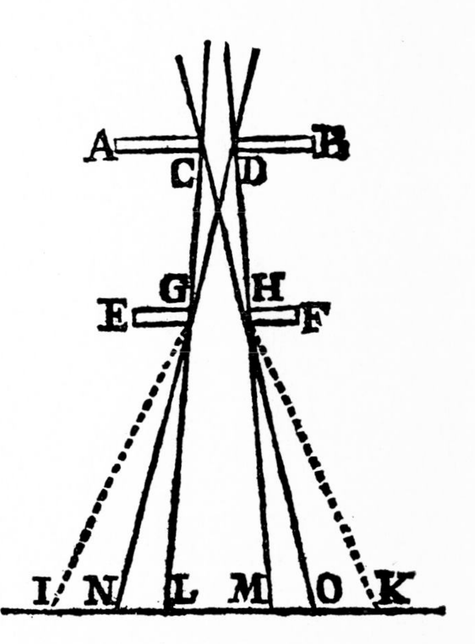

Fig. 49. The second experiment on diffraction, from *De Lumine*

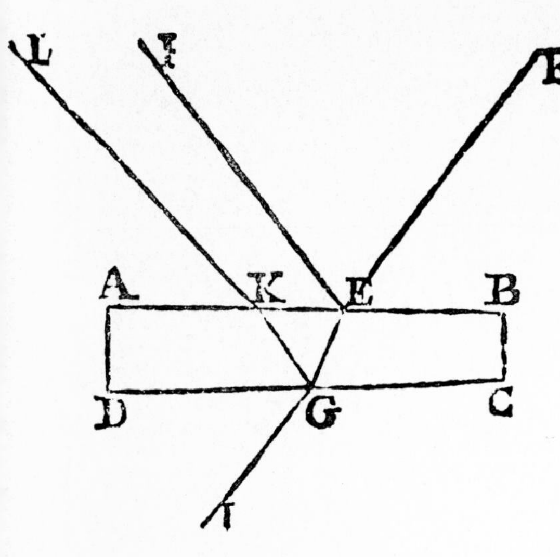

FIG. 50. Reflection from the two faces of a transparent slab, from *De Lumine*

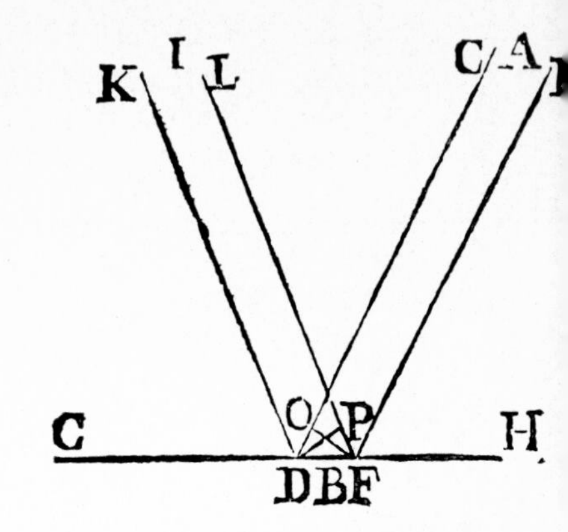

FIG. 51. Diagram showing the reflection of light, according to Grimaldi, from *De Lumine*

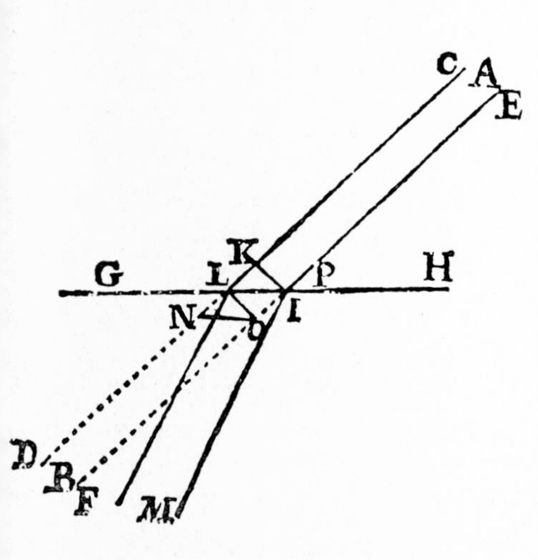

FIG. 52. Diagram illustrating one of the 17th century theories of refraction, from *De Lumine*

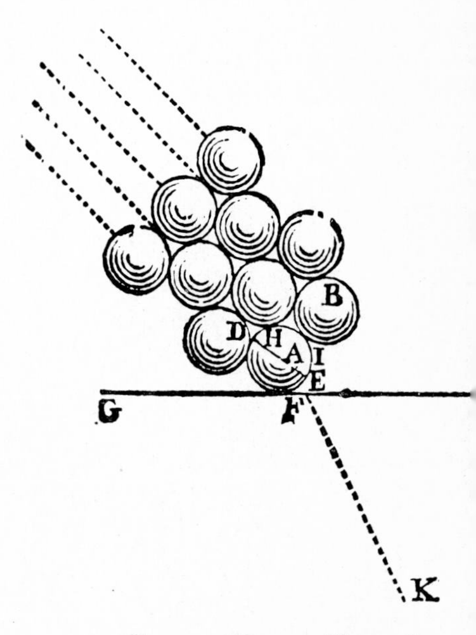

FIG. 53. Diagram illustrating another 17th century theory of refraction, from *De Lumine*

that the permanent colours are something in effect distinct from *lumen* and residing in the bodies as it is generally believed.

We cannot fail to be surprised when faced with an author who writes 472 pages to support one thesis against another, and who then writes 60 pages to conclude that after all, apart from some secondary considerations, the thesis he was attacking could well be right. Was this a case of conversion? It is doubtful. Conversion might have taken place if Book Two had been published separately, and after, Book One; but it is a single work and it is the work of someone who has spent his whole life, unfortunately a short one, in the study of the subject and who, in the end, having decided to summarize the results, found that he had to admit that between the two possible hypotheses he tended to support one but he could not completely exclude the other. This is not conversion but rather perplexity, and this is evident in the Proemio. Grimaldi began by stating the fact that although light existed, as everybody who was not blind knew, its nature and its consistency (*naturam et quiditatem*) was a great mystery.[7] He attacked most of the philosophers who with long discourses and with '*aenigmatica verborum mysteria*' tended only to confuse ideas while asserting that they were making them clear. This was an obvious allusion to the difficult and abstruse discussions of the *perspectivi* and of the Scholastics. Grimaldi showed the sincerity of his intentions at once; he wrote:

> ... let us be honest we do not really know anything about the nature of light and it is dishonest to use big words which are meaningless.

He continued:

> ... nevertheless I do not think that we should call daring the one who attempts to encourage these studies....

This sentence reveals the cultural climate in which Grimaldi was working. As we have mentioned before, in those days the classical texts of the great masters of the past were held to contain the truth

[7] Op. Cit. Proemio

and the whole truth. Whoever attempted to raise doubts and to advance original ideas was immediately accused of being imprudent and this was not a negligible accusation, it was equivalent to being considered a revolutionary. At the beginning of the Proemio, Grimaldi apologised for his approach. As the title of Book One stated explicitly, he defended the opinion 'of some', namely of a minority. He knew that he would have to face the accusations and the acrimony of the ruling circles of uncompromising Peripatetics. Grimaldi defended his work not only by hinting first at 'certain new experiments' which at once placed him in a safe position, but also by accusing the traditional views and studies of being inadequate. He justified his procedure simply by saying that until then much had been said with the sole result of hiding the actual ignorance concerning the nature of light under too many meaningless words. None of these words could explain the new phenomena which he had discovered; according to him therefore, not only was it not imprudent to investigate this field, but it was his duty.

From the beginning of his book Grimaldi stated that he proposed to study the question of the nature of light by developing and interpreting his own experiments without paying any attention to the authority of the Masters. The spirit of the seventeenth century was obviously very alive here and Galileo's example and ideas began to bear fruit. Nevertheless the great changes were not immediate and we detect in Grimaldi's book, along with these good intentions and their consequences, a certain dialectic tendency that was the real inheritance of the recent past. Grimaldi's book deserves a very careful and detailed examination because it is a real mine of information for optics and for physics in general. Although we do not propose to omit anything that is of importance to our subject the following summary has to be brief.

Book One opens with the discovery for which posterity remembers the name of Grimaldi, namely the diffraction of light. The name was given to this phenomenon by Grimaldi himself and is still in use, notwithstanding the attempts made by many physicists and Newton in particular, to change it because it interfered with their theories. The title of the first Proposition is: '*Lumen propagatur seu diffunditur non solum Directe, Refracte, ac*

Reflexe, sed etiam alio quodam Quarto modo, Diffracte.' The description of fundamental experiments followed. He described the fringes that appeared when light, entering through a small hole into a darkened room, grazed very small obstacles and then reached a white screen. In these conditions the shadow was not separated from the illuminated region, neither was there a simple penumbra as it would have been reasonable to assume given that the dimensions of the source were not negligible. Very complex phenomena took place and light seemed to invade the region of the geometrical shadow, while dark bands steeped in the light of the illuminated region, appeared to follow the edge faithfully. Such a discovery was dangerous for anyone who was attempting the study of the nature of light, given the general cultural atmosphere at the beginning of the seventeenth century. The cautious tone of the Proemio is fully justified!

From the technical point of view we must acknowledge that Grimaldi's discovery has great value. It is obvious that he made this discovery purely by chance, since no reasoning at that time could have led either him or anyone else to foresee a phenomenon of this type. Grimaldi was able to make the discovery because he set himself the task of experimenting with extremely narrow beams of light, that is to say originating from very small sources. He himself said: '. . . having made the smallest possible hole AB in a window, let the light of the Sun enter through this hole into a completely darkened room'.[8] Although diffraction is always present and is inseparable from the propagation of light, the effects are only clearly visible when the source of light has extremely small dimensions and even then only when we look very carefully behind minute obstacles. Conspicuous phenomena of diffraction do exist, but they are rare, and difficult to detect and interpret. The size of the source of light may seem to be a detail of secondary importance, almost a curiosity, yet many centuries elapsed before scientists decided to make use of it, and even today it is responsible for some of the difficulties of the language of optics and of the experiments in optics. In reality it is an extension into the field of physics of the subdivision into elements and, as we have remarked in the previous chapter, this subdivision was an important turning point in the attempt to explain the mechanism of ocular vision. This tendency towards the infinitesimal has been, and still is, of fundamental importance for the development of investigations in optics.

Grimaldi carried out two groups of experiments with these small sources of light. In the first he interposed an obstacle in the path of light and observed the shadow cast on a white screen. In the second group of experiments he placed an opaque screen with a small hole in the cone of light, so as to let through a very narrow beam of light, and again he observed the results on a white screen. In the first case Grimaldi observed that light invaded the geometrical shadow and that the region of shadow was larger than it should have been, that is to say larger than that expected from the geometrical projection of the obstacle upon the plane of the screen. In addition, he observed that around the edge of the shadow there were 'three series', namely three luminous bands which were coloured, bluish on the side nearest the shadow and reddish on the side away from the shadow. The intensity and the width of these bands decreased as their distance from the shadow increased. These '*series lucidae*' (and the author insisted that they were '*lucidae*' and coloured, and not shadows) followed faithfully the contours of the region of shadow, even if they were irregular. This phenomenon appeared strange at first, but it became embarrassing when it was observed that in certain circumstances, other '*series lucidae*' similar to the previous ones and also coloured, appeared in the region of shadow, and that they always appeared in pairs, two, four and even six, always symmetrical with reference to the edge of the shadow. While the former series were remarkably constant both in number and in form, the latter series, which were in the region of shadow, followed very complex laws and changed easily both in number and characteristics. (Figs. 45–48.)

In the second group of experiments Grimaldi observed that the image made on the white screen by the thin pencil of light which had passed through the second tiny hole, was much larger than that expected from geometrical projection, so much so as to exclude any possibility of a mistake. The central part of this image was

[8] Op. Cit. Proposition I No. 7

white ('*perfusam mero lumine*' said the author), and the edges were coloured red and also blue. Grimaldi carried out several tests: altering the distance of the white screen from the obstacle or from the hole, changing the shape, the colour, the substance and the position of the obstacle or of the diaphragm. The results were disconcerting because with the changing of the distance of the screen upon which the '*series lucidae*' were observed, Grimaldi found that: '. . . they are neither on a straight line passing through the source and tangent to the obstacle, nor on a straight line joining the hole to the point of the screen on which the '*series*' are observed'.[9] On the other hand the '*series lucidae*' remained unchanged 'whatever might be the opaque body interposed in the cone of light, whether dense or tenuous, whether smooth and polished or rough and irregular, whether hard or soft'.[10] Grimaldi excluded the possibility that the '*series lucidae*' were due to direct light, and it was not difficult to reach this conclusion since the '*series*' behaved so strangely when the distance of the screen was altered, as we have already explained. He also excluded the possibility that they were due to refracted light, because they were not affected by the substance of which the opaque obstacle was made, nor could they be due to reflected light even from the edge of the obstacle since the '*series*' remained unchanged whatever the edge and however the obstacle moved.[11] Grimaldi concluded that light must be propagated in some other peculiar manner for which a name had to be found and he called it 'diffraction'. (Fig. 49.)

In the Second Proposition he attempted to explain the phenomenon and here the complications began. He himself recognized that the phenomenon was mysterious, nevertheless he tried to find an explanation for it. He compared light to a fluid and, as a pebble thrown into water produces waves around the point where the

[9] Op. Cit. Proposition I No. 17. This is an interesting passage because its content was troublesome to the supporters of the corpuscular theory, it was conveniently forgotten for a century and a half.

[10] Op. Cit. Proposition I No. 18.
This passage too was purposely forgotten, as we shall soon see.

[11] This experiment too was rediscovered a century and a half later

pebble strikes the surface of the water, he suggested that something similar must happen in the case of light, almost as if the obstacle interposed in the cone of light acted as the pebble on the water.[12] It is obvious that this theory was out of place. The heading of the Second Proposition is worthy of note: 'It seems that *lumen* is something fluid, capable of passing very rapidly and at least sometimes also as a wave through a transparent body'. Therefore Grimaldi was led to consider right from the beginning that *lumen*, 'at least sometimes' must be propagated through transparent bodies 'as a wave'. Grimaldi carried out many experiments on diffraction with fine wires, with linen cloth, with birds' feathers, etc. (Proposition XXXI), but because he was only interested in showing that in these cases light was coloured, he did not develop these experiments systematically. In the absence of a real theory of the phenomenon, it was too difficult to organize experiments in such a complex field. It was already a great achievement that he had succeeded in dealing radically with the experimental part of the simplest and most typical experiments on the subject.

Before examining other parts of the volume we shall call attention to its controversial form, so complete and so sincere, as to exclude the possibility of being only a work of rhetorical form. The whole work has this form of debate: thesis, demonstration, objections, answers to the objections and sometimes refutation of the objections and rejoinder. It is almost like reading the proceedings of a court trial. There are also violent outburst of anger against the stupidity of those who refused obstinately to believe certain 'theoretical acrobatics'. In the course of one of these discussions we find the following passage:

> Secondly you will say that this radiation of *lumen* is such because the air which is illuminated having become like a new source of light creates its own sphere of activity and as a result a secondary *lumen* is produced in multiple ways and in such a manner as to produce these numerous series.[13]

[12] Op. Cit. Proposition II No. 16.

[13] Op. Cit. Proposition I No. 21

If we suppress the word 'air' which renders a little too material the medium in which light travels, we find in these lines a concept that is the same as the one still used today to explain diffraction; and that was proved nearly a century and a half after Grimaldi. It is curious that Grimaldi placed this idea among the four opposite opinions (it is the second as mentioned in the above quotation), and then demolished it. The arguments are so obvious that no one would dare to say that he is wrong. In fact he remarked that had these secondary spherical activities existed, they would have been less intense than the direct beam of light and hence their action would have been invisible or at least almost invisible on the 'continuous' background produced by direct illumination. On the other hand, if light emitted by these secondary centres did exist, it should have been possible to collect it upon a screen placed at the side of the beam, while in actual fact this did not happen. Finally with reference to these secondary sources, it was not easy to understand why the light was coloured, why it was distributed in several '*series*' and why it was necessary to introduce an obstacle in the beam of light in order to observe the phenomenon. 'All the above arguments show how strongly this objection, namely this answer of the adversaries, should be rejected'. Grimaldi must have worked very hard to refute this objection, and his answers showed an invincible logic for his time. What is wonderful in this as in many other cases, is man's intuition which is often almost prophetic. It is incredible that in the first half of the seventeenth century there should have been a man who with so few assumptions and such limited and sometimes misleading knowledge, attempted to explain the diffraction fringes by a mechanism which was only imposed two centuries later by the development of our knowledge of light. Naturally this prophet was opposed and his name was forgotten.

We have related these facts because apart from their intrinsic interest they also show that diffraction has been one of the most controversial of scientific debates. In a way it was the rock against which the various theories of the greatest minds were shattered; and, at the same time, it was the vehicle by which the knowledge of the nature of light progressed.

Let us return to the discussion of Grimaldi's work. He asserted that light existed, and he proved it not only by means of the phenomenon of vision, which according to him was due to '. . . an agent stimulating the power of vision and received in the eye . . .' but also by using a new argument which he seems to have been the first to use. Light existed because it heated bodies. In Proposition XXXII No. 8 Grimaldi wrote: 'It is certain that *lumen* exists outside the eye, because in bodies which are illuminated we feel also heat as an effect of the *lumen* received by them. . . .' This conclusion was reached after many experiments and long discussions. At first Grimaldi proved that transparent bodies were heated much less than opaque bodies; among the latter, black bodies were heated more than white bodies. He linked this phenomenon with the greater or lesser absorption of light and, still relying on experiments, he asserted that perfectly transparent bodies did not exist, and even those which appeared so were heated a little by light. Moreover, in order to refute the objection that the heating of a piece of iron exposed to the light of the Sun was due to the heat of the medium interposed he mentioned one of his experiments. He placed a glass vessel full of cold running water between the Sun and the piece of iron and he showed that the piece of iron became heated just the same. He remarked that if we wish to warm our hand in the Sun the presence of the Sun is necessary, but the heat on the hand is not produced directly by the Sun, as some of his opponents claimed. The Sun produced the *lumen* and this warmed the hand; in fact, if the *lumen* ceased to arrive, the heating effect also ceased. Nowadays this type of reasoning appears superfluous to us or at least outmoded, but Grimaldi's book is full of reasoning like this and contains discussions of the most obscure details. This goes to prove what strange concepts arrived at by the most extraordinary rationalistic acrobatics were current among the more cultivated people of the period. The followers of the new *philosophia naturalis* found their way barred at every turn by preconceived ideas of the other *philosophia* and to win a victory was a formidable task. Most of this victory was due to Grimaldi. He quoted other worthwhile experiments to show, for example, that *lumen* from different sources such as from flames or from the Sun, had a different heating effect, even if it appeared equally bright. When some philosopher objected that this heating could be a 'quality' of solar light, Grimaldi retorted that he should try to warm himself in lunar light, which after all was reflected solar light.

Having said all this we must not ignore the fact that some explanations were out of place, such as the question relating to the different heating of a body according to whether it was in motion or at rest. Nor should we be surprised if three centuries ago errors were made in this subject; they were due more to a lack of theories than to incorrect experiments. Grimaldi thus asserted his firm conviction of the objective existence of light: 'If it is not evident that it is the *lumen* that illuminates objects, then there will never be any evidence to be sought or to be hoped for'.[14] With this certainty he faced the arduous task of defining the essence of *lumen*, the '*naturam et quiditatem*'. He set the problem in this form: 'Is *lumen* substance or accident?' In order to find an answer he examined the fundamental properties that experiments attributed to *lumen*, such as the propagation in diaphanous bodies, its behaviour with opaque bodies, reflection, refraction, diffraction, colour and vision. The detailed and thorough discussion takes over four hundred pages and the greater part of the observations and arguments are still true today. Grimaldi deserves admiration for the enormous size of his experimental and rational work. Naturally we shall have to limit ourselves to a summary of his work, since it is not our intention to write a book twice the size of that of Grimaldi.

The discussion seems to be conducted rather by a defence counsel than by an unbiased judge. He affirmed on several occasions that the majority of 'Philosophers engaged in more serious questions' did not appear to give enough weight to the observations and experiments which physicists were collecting; and he confirmed that the most widely held opinion was '*Opinio Peripatetica de Luminis Accidentalitate*. Grimaldi took the opposite view, in favour of *Opinio aliquorum de Luminis Substantialitate*. This was the minority opinion and he defended it strongly, showing that

[14] Op. Cit. Proposition XXXVIII No. 5

in all phenomena (propagation of light in diaphanous bodies, reflection from diaphanous or opaque bodies, refraction, colour and vision) the ideas could be explained satisfactorily by admitting that *lumen* was substance, while hundreds of objections would be encountered by accepting that '*lumen* was accident'. From this discussion was born a *lumen* which had the properties of a fluid substance, extremely tenuous, very fast and endowed with local motion. Even he, however, realized that many of the arguments in favour of a '*lumen* substance' were weak and that various criticisms of '*lumen* accident' were not really convincing. In Book Two he stated this opinion with great honesty, and he summarized the reasoning put forward together with the replies in favour of the opposite argument. We do not propose to enter into the details of this discussion; but it will be useful to us when studying the evolution of ideas on this subject in the following centuries to know something about Grimaldi's definition of the characteristics of *lumen*.

The concept of light as 'quality' (*Accidens de genere Qualitatum*) was already subject to so much criticism that there is no need to dwell on it. The concept of light as a substance which was very fluid, very tenuous and very fast met with serious rational objections, the foremost of which was its penetration of diaphanous bodies. The main difficulty concerned the idea of light travelling in a vacuum since in Grimaldi's time vacuum pumps were not in use, and the ruling philosophical concept, 'Nature abhors a vacuum', meant that a vacuum could not exist. To explain transparency, therefore, Grimaldi turned to the concept of porosity of matter. He described many experiments to convince his readers that this quality existed in all bodies. But admitting the existence of pores, or of numerous extremely small channels, was a long way from accepting the permeability of a body to a matter, even as fluid and tenuous as light was considered to be. This permeability had to be 'along a line, if not geometrically, at least physically straight' to allow the passage of light from any direction. Moreover, this theory required that the simultaneous beams did not disturb each other, and that the agitation of the body did not disturb the transmission; and without having to accept that all bodies, diamond included, had a structure that was inconsistent, powdery and very fragile. As usual the opponents did not spare their objections, including experimental objections, and Grimaldi himself quoted the case of gunpowder which when it explodes removes all obstacles both solid and rarefied. What would light then do with its great velocity!

One by one these arguments were discussed by Grimaldi, from them he tried to extract what he could in favour of his argument. He was led to conclude that light did not go through diaphanous material by direct penetration but rather in an indirect manner,[15] and he gave several examples: like wine in water; like the wind, which entering a house by the window, always maintains its prevailing direction no matter how many walls it meets; like water in a sponge; like milk in the breast.[16] In all this he was in agreement with the 'vat of Descartes'. Grimaldi did not deny that some rays were intercepted by matter; in other words, light emerging from a diaphanous body was no longer continuous but had a structure similar to that of water passing through a sieve. But he added that it was enough that the discontinuity should be imperceptible. In fact the perfect diaphanous body did not exist; there was, however, a continuous grading from diaphanous to opaque, and even the most opaque material became transparent when in thin layers. After this Grimaldi explained the cause of hardness and of consistency (*durities et consistentia*). He quoted the example of slaked lime,[17] which became harder as it dried; whereas at first it appeared compact, in its dry state it showed visible pores '. . . and small cavities which can be easily observed, at least by means of a microscope'. In other words he attributed the hardness not to the fact that the matter was compact, but rather to the 'peculiar union of the solid parts'. He also considered heat and quoted many examples such as that of wax, which was opaque when cold and became transparent when heated, and foam which was opaque but became transparent when condensed. Next he turned to magnetism. He attributed substance to the '*effluvium magneticum*', and suggested a complete parallel with light. He observed phenomena of capil-

[15] Op. Cit. Proposition IV
[16] Op. Cit. Proposition VIII No. 69
[17] Op. Cit. Proposition VIII No. 89

larity, and attempted an explanation of these as well. Finally, after discussing the nature of the rainbow (in another part of the book) and having shown that there was a complex luminous trajectory in every drop of water, he deduced arguments in favour of the extreme tenuity of light and of pores. But another objection was raised: if pores could not be empty (because nothing in nature could be empty), with what were they filled? Grimaldi thought that they were filled with '*purus aether*'. He also advanced the hypothesis that they were filled with the same substance which constituted light, in a stationary form; and that this stationary light, which was invisible, cold and undetectable, became mobile when a new light from the Sun or from a luminous body reached it. Here we detect the influence of the 'vat of Descartes'. Though Grimaldi did not mention Descartes, it is clear from his refutation of Descartes' ideas on colour that he knew of his work.

At this point it is interesting to find an hypothesis on the origin of solar light, because the opponents of the theory that light had substance had never thought of raising the question of the fate of the Sun, compelled for thousands of years to emit this luminous matter. Grimaldi contradicted the idea that the Sun must undergo a contraction as a consequence of the continuous emission of material light. He quoted Kepler on this subject, and believed that the Sun since its beginning must have decreased in size. Referring to the earliest measurements of the diameter of the Sun carried out by the Chaldeans, he concluded that from those days to his the diameter of the Sun had decreased at most by a minute of arc, this being the order of uncertainty of the measurements. The emission of light over the period could easily be explained by this contraction, and we could be sure that the Sun would last until the Day of Judgement.

Another observation is interesting from the historical point of view, the title of Proposition X: '*Lumen non propagatur in diaphano cum influxu effectivo partis in partem ipsius luminis*'. This shows at once that the author was definitely opposed to an idea which seemed to be very widespread at that time and which we have already mentioned when discussing one of the explanations of diffraction given by the opponents, namely that light was propagated by secondary spherical actions. This is the basic concept that later was attributed to Huygens. The main reason for Grimaldi's objection was that if this type of propagation were true, then the points reached by light would also become sources of light, and therefore should irradiate all around, that is to say spherically. This was contrary to the evidence given by the experiment of the linear propagation of light. According to Grimaldi, it would have been impossible on this basis to obtain the laws of reflection and of refraction, and this was enough to demolish the theory of the propagation of light by means of secondary spherical actions. This reasoning of Grimaldi was faithfully followed for a century. From an historical point of view it is interesting to find this view expressed by none other than the discoverer of diffraction, that is to say of the phenomenon which can be best explained by the mechanism of propagation of secondary spheres. The discoverer of diffraction himself, in order to confirm his argument with certainty, stated that the trajectory of light was clearly rectilinear, and he did this while describing the experimental conditions which had led him to the discovery of diffraction. This goes to show how difficult it is to remove elementary concepts, or preconceptions, once they have taken root. Yet at that time, the concept of successive propagation by means of secondary sources irradiating spherically, must have been familiar and wide-spread. In the same book by Grimaldi are quoted two examples which have been forgotten, although they are extremely suitable for explaining the concept even from the teaching point of view. The first example was that of a fire which was propagated by its sparks. Each spark started a new fire, circular in form, where it landed, thereby giving the impression that it was the central fire which was propagated. The other example was that of a piece of ice exposed to a narrow solar beam. The part touched by the Sun was the first to melt and then, gradually, the surrounding parts melted, although they were not illuminated by the Sun. In spite of all this, Grimaldi felt sure that he had proved the absurdity of this type of propagation of light. In conclusion, the concepts of Grimaldi on the porous nature of matter and on the nature of light can be summarized in his own words:

> I say that it is not beyond the power of the Omnipotent that a corporeal substance should exist that is porous with continuity as we have explained in the case of diaphonous bodies, and that a corporeal substance should also exist that is tenuous, fluid and which is irradiated by a fast and intense motion as it has been said of light. . . .[18]

Let us now turn to the study of reflection. The mechanism of reflection of a material light was obvious. Grimaldi pointed out that it would be extremely difficult to explain the mechanism of reflection of an 'accidental light'. However, even the first argument presented difficulties. When studying the reflection by smooth and by rough bodies, by light and dark bodies, by white and coloured bodies, Grimaldi noticed that sometimes the light acquired colour while other times it did not, that sometimes it followed the laws of reflection as given by Euclid, and other times it was diffused all around. From the examination of all these cases he was led to the conclusion that sometimes light was reflected by the external surface of a body (as in the case of a bright surface of coloured glass), while at other times it seemed to penetrate slightly into the body before emerging. Meanwhile Grimaldi discovered that the percentage of reflected light increased as the angle of incidence increased. We ought to note that here by angle of incidence we mean the term as used in modern times because in Grimaldi's time the angle of incidence was the angle between the ray of light and the surface of the body.

Grimaldi made two important contributions. The first concerned reflection by a slab of glass which was surrounded by air and had parallel or almost parallel sides. The reflection from the first side was easily explained; the glass was denser than air, had less pores and a percentage of light unable to find a way to enter the glass had to return into the air. But why was it that the light which had penetrated the glass was partly reflected when emerging into the air through the second side? The question became even more difficult to answer when water was substituted for air against the second side of the slab. Water was certainly denser than air and yet the intensity of light reflected by water was less than that of the

[18] Op. Cit. Proposition VIII No. 91

light reflected by air. Grimaldi was indeed perplexed and he tried to explain this difficulty by suggesting that the orifices of the pores of glass did not coincide well with those of air or water. He concluded: 'Whoever will be able to reconcile this, will have eliminated the whole of this important difficulty and by adding *lux* to *lumen* he will fully deserve its clarity because at the moment for most people it is surrounded by the darkness of error'.[19] (Fig. 50.)

The other contribution was Grimaldi's attempt to give a rational demonstration of the law of reflection. He succeeded in this not by considering a single ray and using one of the models of Descartes, but by considering a cylinder of rays. He observed that only in the geometrical conditions postulated by Euclid's law, did the section of the beam of rays remain constant, and that therefore reflection was possible in relation to the frequency of the pores in the medium in which light travelled. This meant that Grimaldi's fluid had to be endowed with a density of its own that was characteristic of the medium in which it travelled; the trajectory had to be fixed in such a manner as to remain constant. This concept was implied rather than expressed; and in any case, it could have been easily invalidated by referring to the behaviour of curved mirrors. Nevertheless it is widely used when dealing with refraction. When dealing with this subject Grimaldi naturally began by calling attention to the serious difficulty of explaining refraction by means of the hypothesis of 'accidental light'. He continued by reviewing the many theories advanced on the basis of the hypothesis of a 'substantial light'. Generally speaking, he found all these theories quite satisfactory to explain the refraction of light when it passed from one medium to a denser medium, because it was possible to understand the bending of the trajectory on account of the increase of resistance to the light in the denser medium. This had already been hinted by Alhazen. What remained inexplicable was why did the light deviate when it passed from a denser medium to one less dense. In the limiting case, if the narrow beams of light met a vacuum on leaving the substance, why should they deviate, instead of continuing along the rectilinear trajectory according to the principle of inertia?

[19] Op. Cit. Proposition III No. 25

Grimaldi reached the conclusion that not one of the few philosophers who had given any thought to this question had obtained a satisfactory answer. (Fig. 51.)

The summary of the theories which followed deserves to be mentioned even if briefly. One of the theories was based on the different velocity which light had in the two media separated by the refracting surface. The reasoning would not have been valid if a single ray had been considered, but we must remember that it referred to a cylinder of rays considered as a whole. In fact, there was no reason why a body launched on a rectilinear path in a viscous medium should deviate in encountering an increase in viscosity, even across a surface of discontinuity. From the point of view of physics it was meaningless to talk of the inclination of an isolated projectile with reference to the above mentioned surface. On the contrary the various rays were not considered independent and in this way there was introduced a concept similar to that of a front perpendicular to the direction of propagation. Today this is known as the wave front, which is one of the fundamental concepts of optics. The demonstration which followed was well known and elementary. When a beam of rays fell obliquely on the surface of separation of two media, it deviated in such a way as to be less inclined to the normal to the surface of separation in the medium in which the velocity of propagation was smaller, in order that the front should still be perpendicular to the ray. However, these 'tendencies', this light which 'prefered' to remain as a whole, did not satisfy the new optics any longer. Although the phenomenon of refraction could be satisfactorily represented from a formal point of view on this basis, the reasoning was not entirely satisfactory, and Grimaldi reported it without approval. There existed '*putantes lucem in minutissimos globulos resolutam esse*', namely those who considered that light was composed of numerous little spheres. They imagined that a ray was a small cylinder containing several spheres and that this also applied to a normal section. When the ray was cut obliquely by a surface of greater resistance the spheres on one side were slowed down before the others, and the ray was encouraged to bend towards the normal to the surface. Grimaldi objected to this artificial model which was not sub-

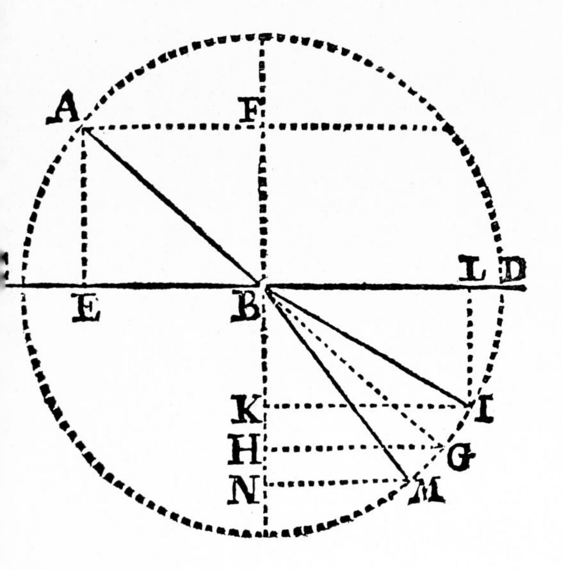

FIG. 54. Diagram illustrating Descartes' theory of refraction, from *De Lumine*

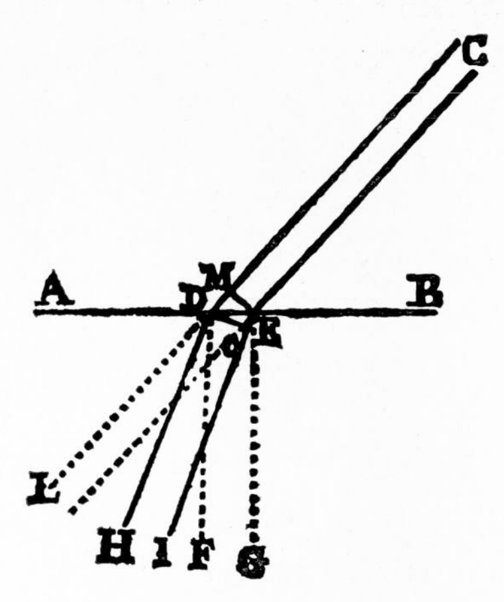

FIG. 55. Diagram illustrating Grimaldi's theory of refraction, from *De Lumine*

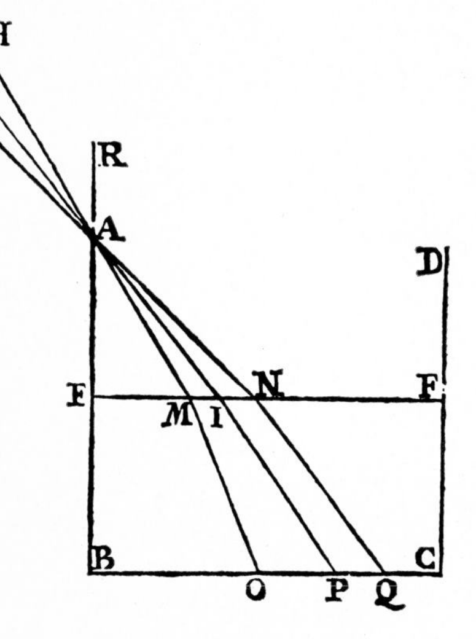

FIG. 56. Experiment showing the colours acquired by light refracted by water, from *De Lumine*

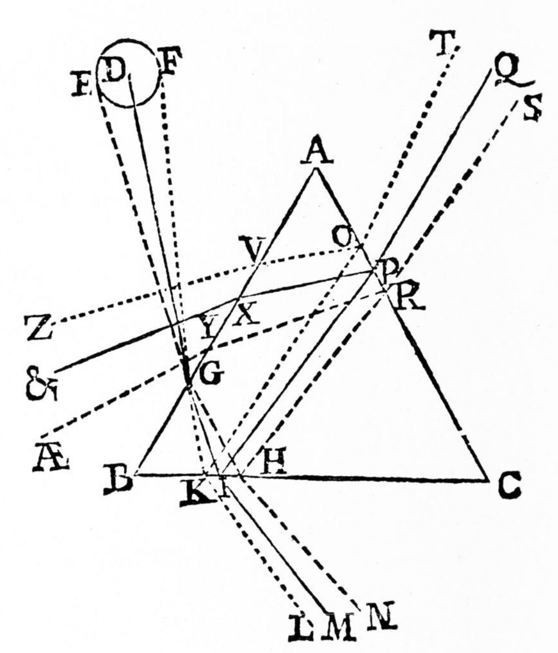

FIG. 57. Refraction through a prism of 60°, from *De Lumine*

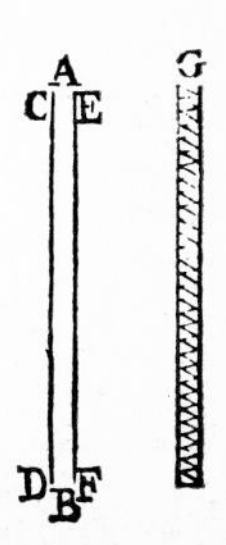

FIG. 58. Diagram of the ray according to Grimaldi's wave theory, from *De Lumine*. (Note that the vibrations are transverse.)

MICROGRAPHIA:

OR SOME

Physiological Descriptions

OF

MINUTE BODIES

MADE BY

MAGNIFYING GLASSES.

WITH

OBSERVATIONS and INQUIRIES thereupon.

By *R. HOOKE*, Fellow of the ROYAL SOCIETY.

Non possis oculo quantum contendere Linceus,
Non tamen idcirco contemnas Lippus inungi. Horat. Ep. lib. 1.

LONDON, Printed by *Jo. Martyn*, and *Ja. Allestry*, Printers to the ROYAL SOCIETY, and are to be sold at their Shop at the *Bell* in *S. Paul's* Church-yard. M DC LX V.

FIG. 59. Frontispiece of *Micrographia* by Hooke

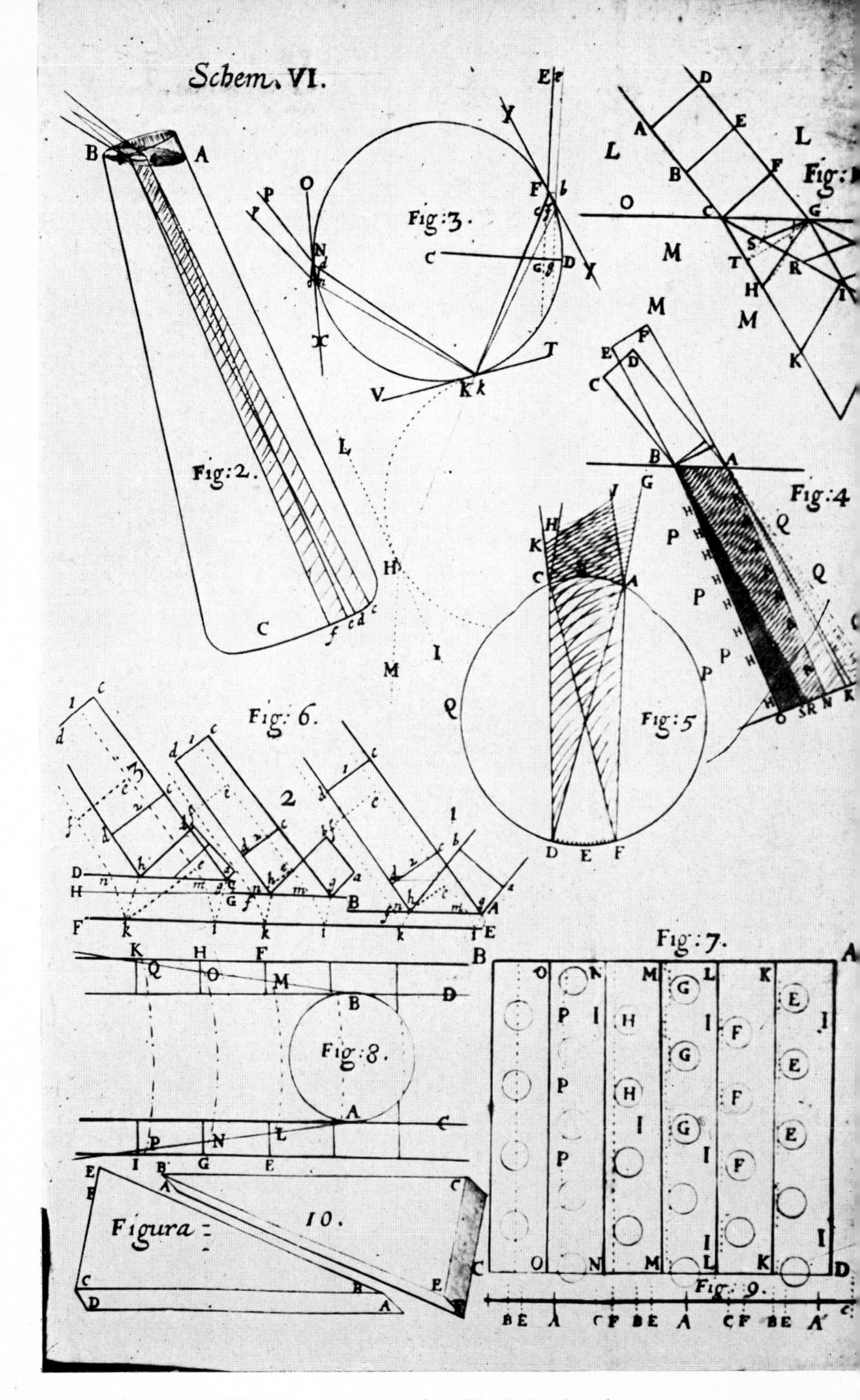

FIG. 60. Diagrams representing Hooke's theories on refraction and on the colour phenomena of thin plates

Fig. 61. Isaac Newton at the age of 84

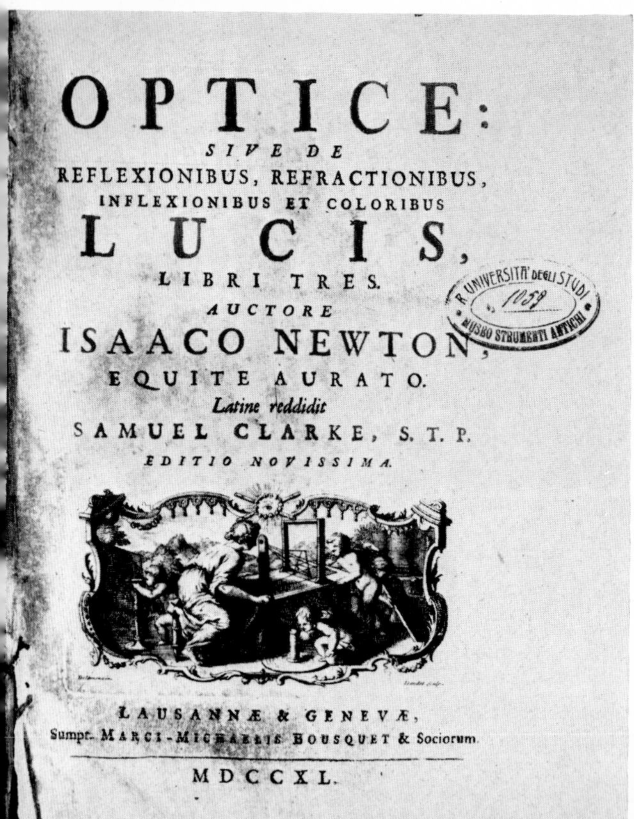

OPTICE:
SIVE DE
REFLEXIONIBUS, REFRACTIONIBUS,
INFLEXIONIBUS ET COLORIBUS
LUCIS,
LIBRI TRES.
AUCTORE
ISAACO NEWTON,
EQUITE AURATO.
Latine reddidit
SAMUEL CLARKE, S. T. P.
EDITIO NOVISSIMA.

LAUSANNÆ & GENEVÆ,
Sumpt. MARCI-MICHAELIS BOUSQUET & Sociorum.
MDCCXL.

Fig. 62. Frontispiece of *Opticks* by Newton

stantially different from the previous one. Some philosophers supported another theory, which did not appear to be very satisfactory. They suggested that the spheres of which light was composed, underwent some sort of rotation when meeting a surface obliquely. (Figs. 52–55.)

Grimaldi also gave the theory of Descartes, though he did not mention him by name. He stressed that this theory required an increase of velocity in the denser medium because the bending towards the normal to the surface necessarily required an increase in the component of motion which was parallel to the normal. Grimaldi did not hesitate to state: 'In reality I consider that this opinion in itself and in its sole exposition appears to me improbable'.[20] Finally having demolished the theories of the others, Grimaldi put forward his own theory. This was based mainly on the concept that the bending towards the normal of a beam of light that was obliquely incident on a refracting surface, implied a widening of the section of the beam itself. This had to occur because of the necessity of compensating, by means of this widening of the section, for the smaller distribution of pores in the denser body. The title of Proposition XX in which Grimaldi expounded this theory is: '*Reddere veram rationem de Refractione Luminis*'. We have the impression that he stated this only to convince himself. We must admit that it was not a very brilliant, nor a very persuasive theory, particularly when we recollect the objections which had already been put forward to a similar theory in the case of reflection. Thus the refraction of light still remained a mystery.

Let us now deal with the question of colour. In all the philosophy of the previous centuries the interpretation of colour presented serious difficulties. Many strange and discordant ideas had been put forward, but all seemed to agree on one point: that light was something quite distinct from colour. Even when light was considered as emerging from the eye, no one ever suggested that colour did as well. It was thought that the light emerged from the eye to seek 'forms and colours'. Alhazen himself always spoke of '*lumen*', of '*lux*' and of '*color illuminatus*'. Colour was a quality of

[20] Op. Cit. Proposition XIX No. 15

the object; when philosophers used the concept of 'husks', they considered that these carried colour with them. When light became an objective entity, the colour remained separated. Light was colourless, as we stated at the beginning of this chapter. At the beginning of the seventeenth century, colour was considered to be something comparable to the 'form' of bodies. The seeing of colours was explained by a mechanism called '*species visuales intentionales*', comparable to the *eidola*, witness some passages of *Margarita philosophica* and the exposition of Descartes. White light, now completely separated, was the vehicle for colours travelling towards the eyes of the observer. We can summarize the position thus: philosophy had ceded light to physics but still retained the possession of colour. This situation did not last long. We have already mentioned the attack of Descartes against the '*species intentionales*'; but his arguments were too weak, they appeared more like analogies for popularising the subject than real theoretical concepts. Much more was required to defeat the strong opposition of the Peripatetics, who, in the seventeenth century, held most of the chairs in the important centres of knowledge. The person to lead the first strong attack on this subject was Grimaldi. He carried out an extensive and systematic programme of experiments, followed by remarkable theoretical discussions, which would be extremely interesting to report in full but unfortunately we cannot do it here. The discussion centred on arguments which we of the twentieth century would never consider, because from our early studies we have been confronted with a model of light and of colour that has appeared completely satisfactory. Nevertheless it is worth remembering how hard it was to reach such a satisfactory model.

Grimaldi set himself the task of studying the '*Lumen non coloratum*', and tried to find the necessary process to make light coloured. As a first step he showed that light could be coloured purely by reflection, without the intervention of a change of the diaphanous medium and without the need of refraction. The experiments concerned were so elementary that we need not discuss them here. He then proceeded to show that reflection was not essential for colouring light, since light could also be coloured, simply by refraction. The experiments related to this are well known to us all, but they deserve some comment. The first experiment consisted of letting a beam of sunlight fall on a white plane surface at the bottom of a vessel filled with water. In this way light was refracted only once at the upper surface of the liquid. Grimaldi noticed two colours at the edges of the beam, namely red and blue, and the latter was more conspicuous than the former. In another experiment Grimaldi let a beam of sunlight travel through an equilateral prism having an angle of 60°. As on other occasions, he did not make use of a very narrow pencil of light, which would have given better results. It is strange that throughout the history of optics we find this reluctance to reduce light sources and light beams to a small size. As we have already seen, Grimaldi owed his discovery of diffraction to using very small holes to let the light of the Sun enter his laboratory; yet he never afterwards attempted to stop down or to reduce the beam of sunlight which had an angular width of 32 minutes of arc, this being the angle subtended by the diameter of the Sun. He did it once and he discovered diffraction. Had he continued to make use of narrow pencils of light in this way who knows what else he would have been able to discover! On the other hand, in the experiment with the prism, Grimaldi used the light of the Sun which he limited by means of only one little hole. Obviously the angular magnitude of the beam masked many effects which no doubt he would otherwise have noticed. In the actual conditions in which he was working, he could only detect a red colouring at one end of the beam, and blue colouring at the other, with a yellowish colour in between. Continuing his investigations into the nature of colour, he showed that light could acquire colour without the help of reflection and refraction, or the change of the diaphanous medium, but simply by diffraction. He related several experiments carried out with reticular structures, with holes and with slits. He showed that the fringes thus obtained were coloured, without the intervention of reflection, of refraction or of a change of substance. Therefore he concluded that, if the light became coloured by any of these means and yet none was indispensable, then the change had to occur in the light itself, since this was the only thing common to all the experiments. He stated this in the title of his Proposition XXXII: '*Lumen per solam aliquam ipsius modificationem intrinsecam et nulla alia entitate co-assumpta, transit aliquando in Colorem ut aiunt, Apparentem*'. Having established this very important first point, Grimaldi proceeded to define '*aliquam ... modificationem intrinsecam*'. He advanced a theory which appeared neither convincing nor complete, but which nevertheless contained some elements which once again deserve to be considered almost as prophecies, so close is their similarity with present day ideas. (Figs. 56 and 57.)

Grimaldi carried out many different experiments, with reflecting and refracting slabs of glass, prismatic or with parallel sides, and also with lenses, and he studied what today we would call chromatic aberration. He discovered that an observation made by others, whose names he did not mention, was almost always true, namely that the colouring of the light was always accompanied by a change in the 'condensation' of the beam of light. Thus where the beam was denser the red colour appeared, and where it was more rarefied the blue colour was apparent. He remarked that beams of light reflected by spherical mirrors, both concave and convex, became sometimes denser or sometimes more rarefied, but they neither changed nor lost their colour. In addition, he did another interesting experiment, although the result was badly interpreted. The light of the Sun was allowed to fall on a 60° prism, but this time a coloured glass was interposed. Grimaldi noticed that, after refraction, the same group of colours were visible, as when white light was used, but they were a little less bright. The idea behind this type of experiment was good, unfortunately the experimenter was unlucky. Coloured glass lets through a large region of the spectrum and only reduces the intensity of some parts of the spectrum with reference to others. Moreover Grimaldi stated that he had chosen for his experiment a glass that did not give directly the colours that he attributed to the action of the prism, namely red, yellow and blue, which can be obtained with good filters. This unfortunate choice was responsible for delay in reaching certain conclusions; but Grimaldi from the sum total of his experiments reached the conclusion that is expressed in the title of Proposition XLIII: 'The modification of light by means of which it becomes permanently and apparently coloured, or rather it becomes perceptible with the quality of colour, can probably be said to be due to an undulation slightly varying, like a quivering of diffusion, with a certain very fine oscillation, by means of which the light stimulates the organ of vision with a peculiar and given activity.'[21]

Often in a scientific book it is only at the end that we understand what the author meant in the first pages. This is true for the books of Euclid and of Alhazen and also for Grimaldi's, as well as for others that we shall be studying later on. Now we find the reason for the sentence already quoted in connection with Proposition II, where the author suggested that light in a diaphanous body was propagated '*etiam undulatim*'. The whole work pivots upon this undulatory theory which comes to the surface only at the end of the book, in order to explain the nature of colour. The arguments that Grimaldi put forward to support his idea are extremely interesting. He gave two examples to show that the great variety of possible colours was due to as many different undulations. First he mentioned handwriting to show that, though intentionally equal, an infinite variety of letters always different one from the other are produced. Secondly he mentioned sound, which can give as many varieties of tone as there are varieties of undulations. He further showed that on this basis it was possible to give a satisfactory explanation of the experiments of reflection, of refraction and of diffraction, thereby approaching the conclusions which were to be reached in the following century. (Fig. 58.)

Grimaldi also devoted some time to the study of colours some of which he defined as being simple, and others as a combination of the former. From the way he expressed his ideas, we deduce that at that time the 'theory of colours' was also a common subject of discussion. This is not surprising since the art of painting had existed for so many centuries. What is interesting is that simple colours had already been reduced to three red, yellow and blue. Thus the trichromatic theory was born.

Reverting to Grimaldi's undulatory theory of light, we find that he hinted at the problem of measuring the frequency of vibrations,

[21] Op. Cit. Proposition XLIII

but that he did not attempt it because of obvious difficulties. On the other hand he insisted on defending his concept concerning the nature of colour, a concept that did not appear to be well received. Here again he mentioned the theory of Descartes and showed its weaknesses. Grimaldi did not spare reproaches, and indeed insults, when dealing with his opponents, and this seems to indicate that the controversy must have been particularly violent. Some quotations are relevant here, because they help to highlight the extremely valuable work of Grimaldi in this phase of the struggle for progress towards a goal which we, nowadays, are apt to accept as easy and natural. With the sub-title: '*Subvertitur principale fundamentum Opinantium in contrarium*' Grimaldi wrote in No. 39 of Proposition XLV:

> What has been said so far seemed to be quite sufficient to demonstrate the truth of our Proposition. But in discussing several times this question with men of great learning, I discovered that more than once they made recourse to the evidence of the eyes which according to them revealed colour as being something permanently attached to bodies which are visible and not self luminous, and as something really different from light; and it was not possible by making use of any argument whatsoever to move them from this sacred altar to which they were clinging. [He related the arguments for and against and finally burst out with: '*Euge. Praeclaram sane philosophiam*, *quae quoties mihi decantatur totie irritatem bilem.*'[22]]

Grimaldi had also to defend himself against those philosophers who accused him of tending by these theories to transform vision into an operation of touch. On this subject he quoted the case of a man who in the presence of some high dignitaries had proved that he could distinguish colours by means of touch, while blindfolded and with his head turned away, and added that there had been an official record of this demonstration.

To quote all the observations and all the remarks contained in Grimaldi's book would take too long. We shall only mention here that although he did not know the velocity of light, he showed that light must have a velocity; in other words, he showed that the

[22] Op. Cit. Proposition XLV No. 41

velocity of light could not be infinite. He suggested the possibility of measuring this quantity over a distance of thirty or forty miles, arranging the experiment in such a way that light would travel this distance twice. There is no doubt that Grimaldi left the subject of light much more advanced and clearer than it was before he started his studies. Light was no longer a colourless 'something', which could undergo linear propagation, reflection and refraction, and was capable of stimulating the vision of illuminated objects. It was a material fluid endowed with very high velocity, but not infinite; that travelled and was propagated by means of rectilinear rays, which were not simply defined by their direction but were endowed with an undulation of very high frequency that enabled them to stimulate the sensation of colour. This material fluid could be reflected following Euclid's laws, but with modifications in direction, intensity and undulation, according to the nature of the reflecting body. It could be refracted with similar complications; it could be diffracted by passing around thin objects or through narrow openings; it could heat bodies when it was absorbed by them, and it could pass through bodies when not heating them; it had a close analogy with magnetic flux. Grimaldi realized that the material nature of light had not been proved beyond doubt, and he clearly stated this by also putting forward the reasons which could justify the 'accidental nature' of light. Nevertheless he insisted that even with this new form of light, his ideas on the nature of colour remained unchanged. It was no longer possible, he thought, to assert the general belief that colour existed within bodies and was something distinct from light.

Grimaldi died on December 28th, 1663, at the age of 45, just after completing the book which we have discussed, which was published in 1665. At that time Huygens was 36 and Newton was 23. Grimaldi's book is a very precious work, not only for its technical content, of which the most important is the discovery of diffraction, but also from the historical point of view. In it we find a valuable documentation of the general condition of culture and of the prevailing ideas in the fields of the physics of matter and of light, at a time when other great scientists were making their contributions to these subjects. In Grimaldi's work we find evidence of the limitations imposed upon writers by the authorities of the time. An obvious indication of this is the almost complete absence of quotations, the only exception being Aristotle. As Grimaldi could not quote the authors who had contributed so much to the subject, either because they were protestants or moslems, he followed an impartial line by not quoting anyone.

There is, however, another interesting point. Grimaldi's book was published sixty-one years after the *Paralipomena* of Kepler, and in it we find nothing relating to this. Not only were Kepler's new ideas not accepted, but Grimaldi devoted a very long Proposition, the 40th in Book One, to the study of vision, without making any reference to the retinal image, although he discussed the question whether the *species* were necessary or not to explain vision. This would seem to indicate that after sixty years Kepler's ideas were forgotten, and would seem to imply that not only had they not been appreciated by the mathematicians and philosophers of the time, but that they had not even been considered worthy of study. Since these ideas were new and philosophically in opposition to the concepts concerning vision which dominated the classical texts, they were attacked with the most powerful weapon available in such cases, namely a conspiracy of silence.

Grimaldi devoted the whole of his short life to teaching in the Jesuit College in Bologna. The book he had finished just before his fatal illness was left in the care of Father Giovan Battista Riccioli, a well-known astronomer with whom Grimaldi had collaborated in many observations. It must not be thought from what we have said, that the opposition to Kepler's ideas concerning the mechanism of vision was limited to the circles subject to religious authority. Similar limitations were found in most publications of the time, such as *La Lumière by* De la Chambre, whom we mentioned earlier, when discussing Fermat's work. De la Chambre was a very eminent person in Paris, he was *Conseiller du Roy en ses Conseils et son premier Médicin Ordinaire*, as is stated in the frontispiece of his work. This book was dedicated to Cardinal Mazarin, had the special sanction of the King and was published by Jacques d'Allin in Paris in 1662. It is therefore a work which received official approval and in it, when dealing with vision it is stated: '. . . from which we must conclude that either light produces these images and carries them with it in the medium and to the eyes, or that it finds them ready made and enables them to be perceived'.[23] The 'images' to which De la Chambre referred are in other parts of the book called also '*espèces*', therefore we are still at the same stage as Della Porta. Every trace is lost of the new optics, that is to say the optics which has dominated until the present day. It is not surprising, therefore, that the majority of students today, even those interested in optics, do not know its origin. The evolution of ideas in the new philosophical climate produced by the great revolution of the seventeenth century, was beginning to assert itself, and before long it produced remarkable results.

In 1665 there appeared a new book of great importance: *Micrographia* by Robert Hooke (1635–1703). There is little doubt that this is a crucial work in the history of optical instruments and of the microscope in particular. It is full of observations, carefully and skilfully made with the microscope, which was then only at the beginning of its great career. From our point of view, this work is of interest because it contained the first systematic studies of the important phenomenon of the colour of thin plates. During his observations with the microscope, Hooke turned his attention to the very thin plates, or films which were produced on steel and on metals in general when they were red hot. He also studied soap bubbles, and glass blown into extremely fine sheets, and he noted their colouring. From a systematic investigation of the phenomenon he deduced the following:

1. The colouring appeared whenever a very thin layer of a transparent body was limited by reflecting bodies with different refraction from that of the material forming the layer.

2. It was not necessary that the bodies limiting the layer should both be of the same material.

3. It was not necessary that the layer should be of uniform thickness; if it was then the colouring was also uniform. If the

[23] De la Chambre, *La Lumière*. Jacques d'Allin, Paris 1662 p. 252

thickness varied, as in the case of a lens, then the colouring took the shape of concentric rings of different colour; starting from the centre the colours were red, yellow, green and violet.

4. In order that the colouring should be visible it was necessary for the thickness of the layer to be between a maximum and a minimum, if the thickness was too great then the illumination was white and uniform.

5. There was no need of any special light in order to produce the phenomenon, nor did it depend on the intensity or the position of the source of light.

6. A sudden change from shadow to light was not necessary to produce the colouring of the thin plates or layers. (Fig. 60.)

It is wonderful how Hooke succeeded in reaching and collecting together so many conclusions. He described some of his experiments in the few pages which he allotted to this subject. He added that the presence of the layer or film was hypothetical, since he never succeeded in seeing or measuring it, not even with the most powerful of his microscopes. Yet most of his conclusions can be confirmed today. He used these conclusions to demolish the hypothesis of Descartes, according to which light consisted of little globules, and that colours were due to the rotation of such globules. Descartes had predicted that this rotation began with refraction, and that it should not be present when light already refracted underwent a second refraction opposite to the first, as was the case of refraction through slabs of glass with parallel sides. Hooke pointed out that in the case of the thin plates that he had studied, these were exactly the conditions, and yet the colouring still remained. The sixth of Hooke's conclusions, which at first glance may appear strange today, also contained a refutation of the ideas of Descartes. In those days experiments were carried out with beams of light having a rather wide angle, and Descartes, as well as others, had noticed that the colouring always appeared when passing from a bright zone to a zone of shadow. Hooke showed that in the case of thin plates this condition was not necessary. All this led Hooke to consider that the ideas of Descartes had been overthrown, and he went on to advance a new theory. This had no great value, but deserves to be mentioned as it helps us to understand better the period of time with which we are dealing; a preliminary and preparatory period in which suggestions were put forward that did not last long, but which in some ways prepared the ground for the theories that were to triumph. According to Hooke, light was due to a motion of matter, a motion which had to consist of vibrations because had it been otherwise it would have led to the disintegration of the luminous object. As a crucial experiment, which he attributed to Clayton, he quoted the case of a diamond which became luminous when rubbed, hit and heated, and remained luminous even after the disturbing action had ceased, for as long as the movement produced by these actions lasted. For the propagation of light in diaphanous bodies he postulated the existence of bodies susceptible to this motion. This motion had a very high but not infinite velocity (in this as well he opposed Descartes) which was rectilinear and followed radii of spheres, just like ripples on water hit by a stone. Along these lines, pulses (*orbicular pulse*) were propagated, which in the case of white light were transverse, while under the effect of refraction had to assume a given obliquity with reference to the radius. On the other hand when light penetrated bodies (and here Hooke agreed with Descartes, that in denser bodies the velocity of propagation was greater), the 'preceding' part of the pulse which had acquired a given obliquity became 'confused' and weak. From this followed the definition of colour:

> Blue is an impression on the Retina of an oblique and confus'd pulse of light, whose weakest part precedes and whose strongest follows. Red is an impression on the Retina of an oblique and confus'd pulse of light, whose strongest part precedes, and whose weakest part follows.[24]

This was an ingenious idea! Hooke made use of this model to explain the colours of thin plates: the combination of the pulse reflected by the lower face with that reflected by the upper face gave a combined pulse which was either red or blue or any intermediate colour. He concluded Observation IX of his book with these words:

[24] Hooke, *Micrographia* 1665. Observation IX, p. 64

Thus have I, with as much brevity as I was able, endeavoured to explicate (*Hypothetically* at least) the causes of the *Phaenomena* I formerly recited, on the consideration of which I have been the more particular.

First, because I think these I have newly given are capable of explicating all the *Phaenomena* of colours, not only of those appearing in the *Prisme*, Water-drop, or Rainbow, and in *laminated* or plated bodies, but of all that are in the world, whether they be fluid of solid bodies, whether in thick or thin, whether transparent, or seemingly opacous. . . .[25]

He must have been proud of his theory, and he was not alone in appreciating it since he received praise also from Huygens. Unfortunately this theory did not have much success. However, we must admit that in Hooke's reasoning there was the seed of an idea, which was to be taken up again one hundred and fifty years later; after further development it became a fundamental phenomenon of optics. Hooke was a prophet when he foresaw that phenomena such as the colour of thin plates would lead the way in the formulation of theories concerning the nature of light.

We have dealt at length with the work of Descartes, Grimaldi and Hooke because there is no doubt that their ideas had a decisive influence on the development of the history of light. Their writings reflected the fervour of the argument over the nature of light at that time, and revealed the existence of a general perplexity. Descartes combined the most discordant hypotheses with great serenity; Grimaldi, after long investigations and analyses, was unable to find any decisive argument in favour of *substantia* as opposed to *accidens*; Hooke, although declaring himself ready to explain 'all that are (colours) in the world', found the subject of so little importance that he tried to expound it, as he said, '. . . with as much brevity as I was able . . .' Perhaps we can be forgiven if we think that, had he been convinced that he had solved the question of the nature of light, he would have written a few more pages on the subject, if not a whole volume. However, colours were no longer regarded as '*species visuales intentionales*', but as modifications of light. The law of refraction found a form which we can call final,

[25] Op. Cit. Observation IX, p. 67

and became a formidable weapon in the hands of students of geometrical optics, and experimenters. Diffraction was known; and the first phenomena of interference began to stake their claim upon the nature of light.

Controversy still raged. Among those who took part in the discussion were Galileo whose *Saggiatore* was published in 1619; Marco Antonio De Dominis, whom we have already mentioned; Marcus Marci of Kronland (1595–1667) with his work entitled *Thaumantias Iris, liber de arcu coelesti, deque colorum apparentium natura, ortu et causis*, published in 1648; Isaac Voss of Leyden (1618–1689) who published *De lucis natura et proprietate*; and many others less well known. They all contributed their knowledge, their experience and sometimes their fantasies, in an attempt to solve this complex mystery of nature.

In 1669 a new discovery was made which added to the great perplexity already existing in the minds of the scholars studying the question of light. In spite of Grimaldi's claim in Proposition XX of Book One, of having given the 'true' explanation, it was still far from clear why light passing from one medium to another of different 'density' or 'transparency' deviated from its original trajectory. That the problem was still unsolved can be gauged by the length and the details of the discussions reported, since generally, when a problem is satisfactorily solved, only the final solution is given. While efforts were being made to understand and to explain the nature of light, a Danish naturalist, Erasmus Bartholinus (1625–1698), discovered the phenomenon of double refraction produced by calcite or Iceland spar, and he published his discovery in 1669 under the title *Experimenta crystalli islandici disdiaclastici quibus mira et insolita refractio detegitur*.

When a ray of light is refracted by a crystal of calcite it produces two refracted rays. One, called the ordinary ray, follows the well-known law of refraction, while the other, called the extraordinary ray, follows a different and more complicated law. As we shall see later, this discovery was to have important consequences.

A few years later, in 1676, there took place one of the most important events of the history of light, namely the measurement of the velocity of light. It was another Danish scientist, Olaf Roemer[26] the son-in-law of Bartholinus, who succeeded in carrying out the measurement.

The velocity of light had been one of the most debated subjects in the past, especially since Alkindi, Alhazen and others of the Arab School had insisted that to enable the human eye to see, the existence of an external 'something' of a physical nature was necessary. Two opposing theories existed: one suggested that this 'something' was endowed with a very high but finite velocity, while the other maintained that the velocity was infinite. The failure of every attempt made to measure this velocity strengthened the faction that held the view that the velocity was infinite. It is true that in most cases the reason for believing that the velocity of light was infinite was dictated by metaphysical considerations and often by observations which were both superficial and wrongly interpreted. On the other hand we must recognize that there was a great confusion of ideas. One group of scholars thought in terms of the velocity of visual rays, and the fact that as soon as they opened their eyes they could see extremely distant objects such as the stars, seemed to justify their conclusion that rays had an infinite velocity. Another group thought in terms of the velocity of the *species*, and repeated the same reasoning as that used for the visual rays without realizing that this reasoning, when applied to the *species*, was not logical. While yet another group thought in terms of the velocity of light. Obviously this question of the velocity of light was another of the great mysteries linked with light.

Many doubts still existed even in the seventeenth century. To appreciate this we have only to consider the reasoning of Descartes. As far as the question of the nature of light is concerned, we have

[26] Olaf Roemer was born at Aarhus (Jutland) in 1644 and died in Copenhagen in 1710. He studied and worked in Paris and was appointed by Louis XIV as tutor to the Dauphin. The observations which led to the measurement of the velocity of light were made at the Paris Astronomical Observatory, which at the time, was under the direction of Giovan Domenico Cassini. In 1685, when Louis XIV revoked the Edict of Nantes, Roemer, being a protestant, had to leave Paris and continued his work in Copenhagen.

already called attention to the fact that in the writings of Descartes there is not much coherence or order. But when we come to consider the question of the velocity of light, we discover that his ideas are particularly contradictory. Descartes was always a staunch supporter of the idea that the velocity of light was infinite, so it is difficult to understand how he could have extended the law of refraction, which had been studied and proved in the case of projectiles, to the case of light. On the other hand Galileo suggested a method of determining the velocity of light by means of luminous signals transmitted over a distance of a few dozen miles, although he was convinced that it was a hopeless task to try to discover the nature of light, because of the extreme complexity of the question. Naturally the result could only be negative because of the inadequacy of the experimental means available at the time. Now the history of light takes a new turn.

In August 1675, Cassini discovered a *seconde inégalité* in the motion of Jupiter's satellites, the *première inégalité* being that when the motion was considered with reference to the Sun. Cassini in discussing this question wrote 'this *seconde inégalité* seems to be due to the fact that the light takes some time to reach us from the satellite, and that it takes from ten to eleven minutes to cover a distance equal to the semi-diameter of the terrestrial orbit'.[27] Thus Cassini could be considered as the originator of the method by which the first measurement of the velocity of light was carried out, had he not changed his mind, and after having given this brilliant explanation of the *seconde inégalité*, unaccountably rejected it.

Fortunately this question was taken up again by Roemer. In September 1676, by means of calculations based on the idea advanced by Cassini, he announced that the eclipse of the first of Jupiter's satellites that had been predicted for a certain time on November 9th, would actually take place ten minutes later. Observations carried out on that occasion confirmed the time predicted by Roemer. The velocity of light was now a quantity which had at last been measured.

On November 21st, 1676, at a meeting of the *Académie des*

[27] Montucla, *Histoire des math mathiques* Vol. II p. 580

Sciences, Roemer described his observations and gave the results of his measurements. The news was published on December 7th in the *Journal de Sçavants*[28] and the following year it appeared in English in the *Philosophical Transactions*. [29]As could be expected, this news met with very strong opposition. Incredible as it may appear, the strongest and most hostile opponents were Cassini and Descartes, whose opposition, on account of their reputation, carried some weight. Today the hostility and the objections are forgotten, and the measurement of the velocity of light carried out by Roemer in 1676 remains one of the greatest conquests in the field of optics.

The *Thaumantias* by Marci was published in Prague in 1648. It is a book of a traditional character in which the new ideas already advanced by many scholars, among whom were Kepler and Descartes, do not seem to receive the attention they deserve. Nevertheless in this book, there are some new ideas which are of particular value.

As the full title suggests the work deals with the phenomenon of the rainbow, but it was necessary to advance some theory on the 'apparent' colours to explain this phenomenon. This Marci proceeded to do using mediaeval theories. Marci described many experiments which he carried out with triangular prisms. It was remarkable that in spite of the difficulties created by the concepts he used to explain colours, he succeeded in expressing in a very explicit manner the correspondence between colour and refraction. As far as we know, this was the first time that such a relationship was expressed explicitly. Because of the historical importance of this it is worth giving some further details.

On page 99 of *Thaumantias*, Theorem XVIII stated:

> It is not possible either to have the same colour with a different refraction, or different colours with the same refraction.[30]

[28] See I. B. Cohen, *Roemer and the first determination of the velocity of light* (1676). Isis, XXXI, 1940, pp. 327–79

[29] *Philosophical Transactions* XII (1677), pp. 893–4

[30] Marcus Marci, *Thaumantia Liber* Prague 1648, p. 69

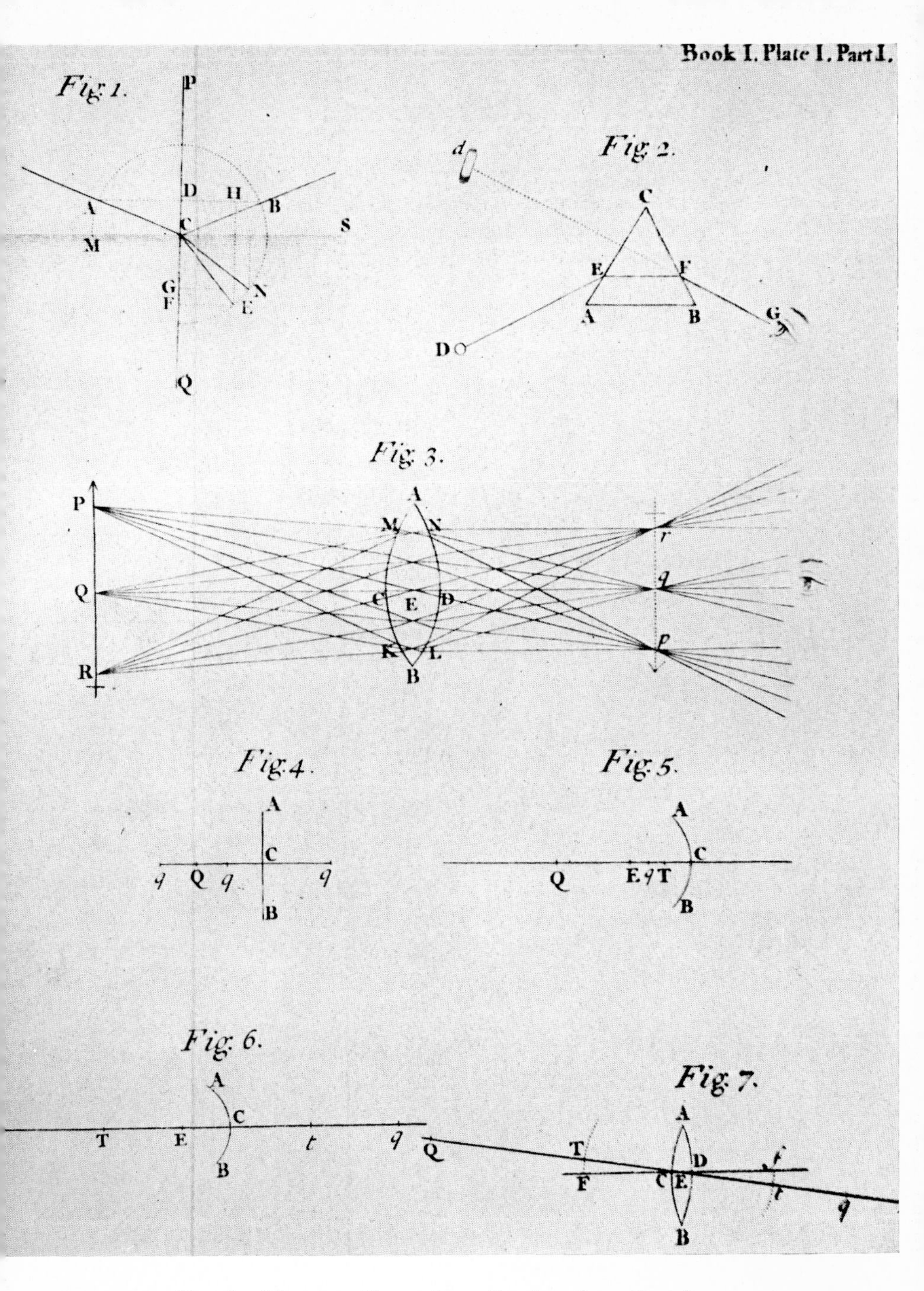

FIG. 63. Diagrams illustrating refraction, from *Opticks*, 1704

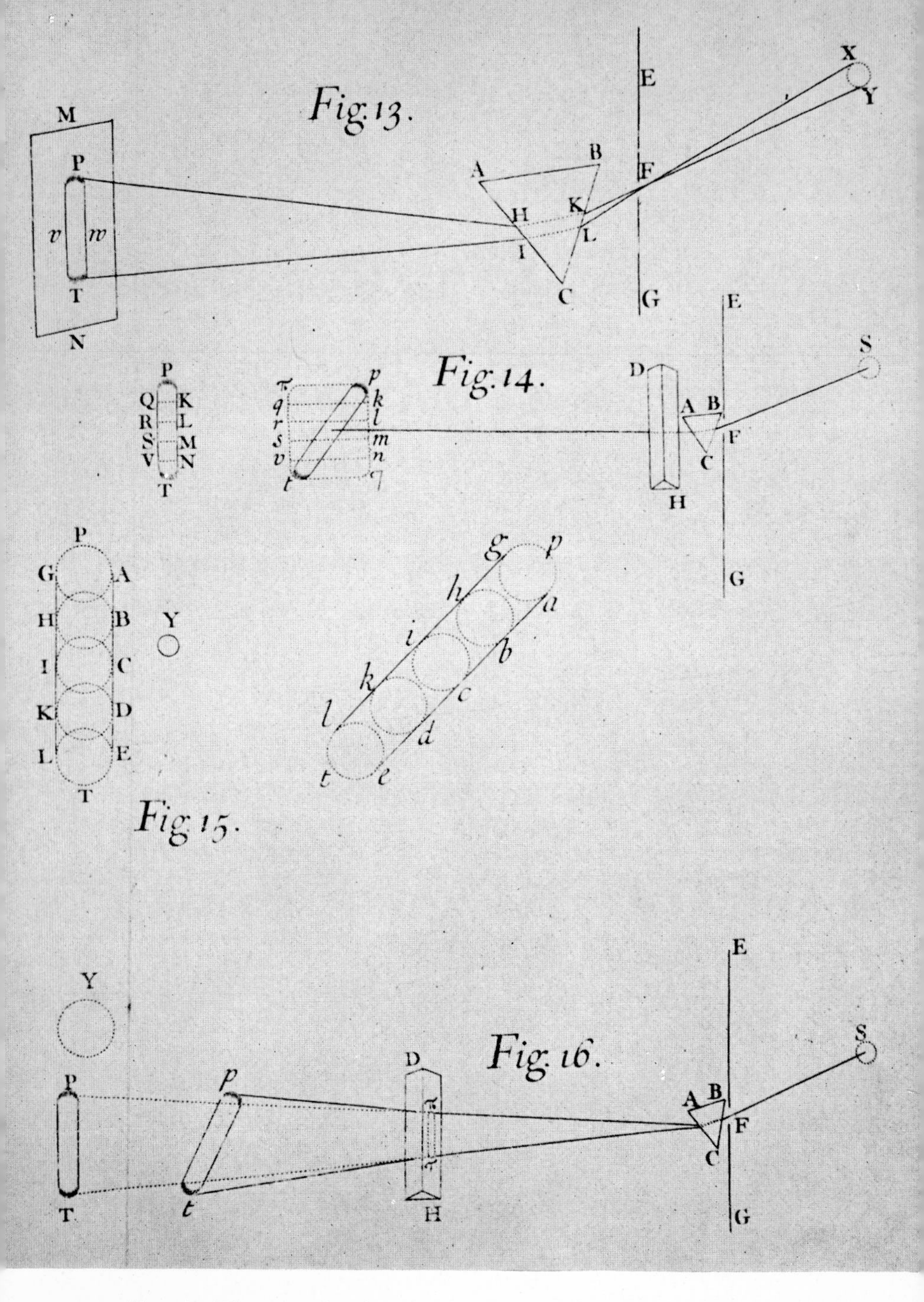

Fig. 64. Diagrams showing the experiments with crossed prisms, from *Opticks*

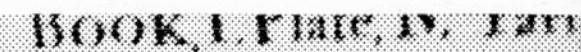

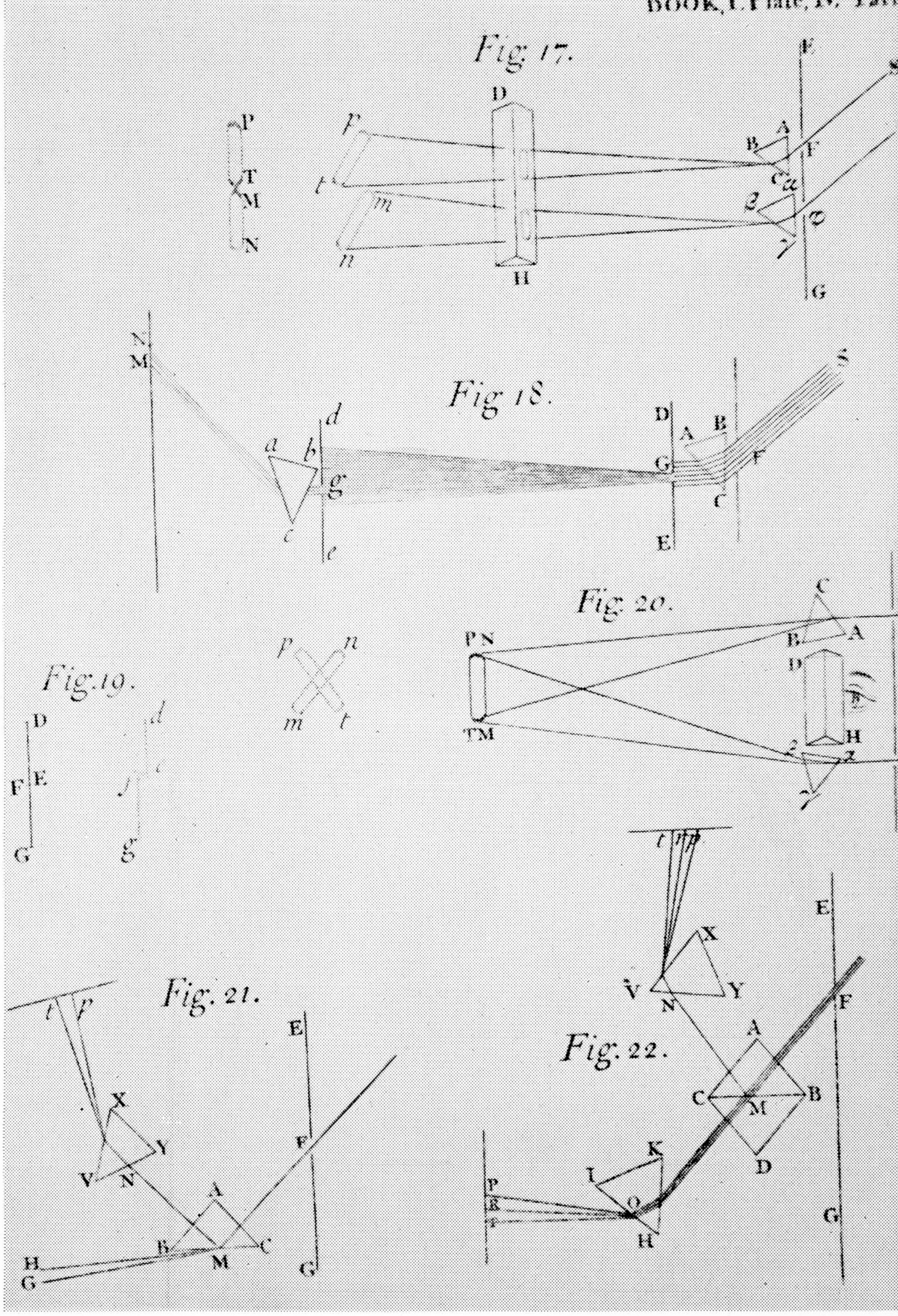

FIG. 65. Diagrams showing the experiments with parallel prisms, from *Opticks*

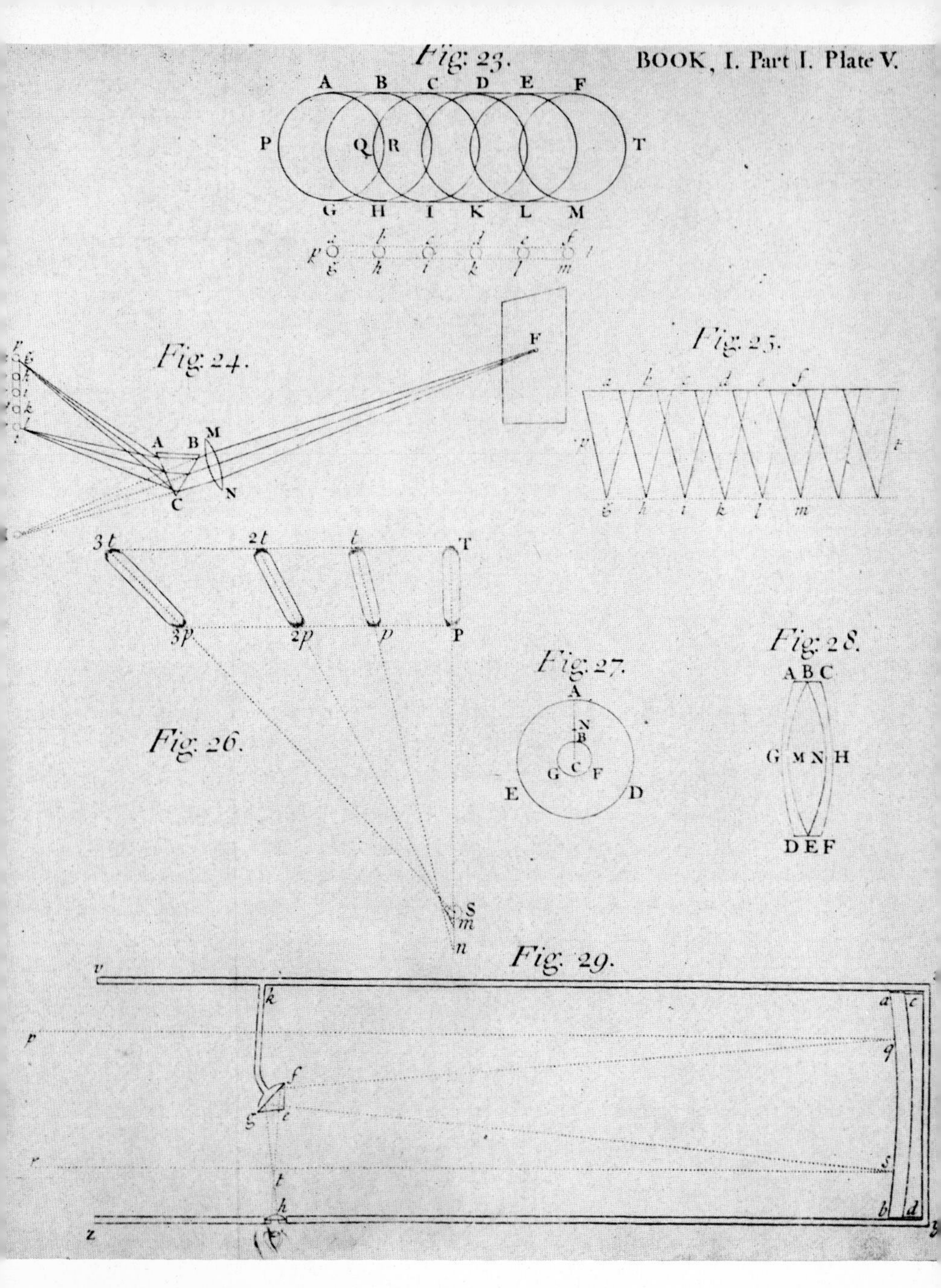

FIG. 66. Diagrams showing the anomalies in the experiment of crossed prisms, from *Opticks*

There is no doubt that this pronouncement contained an assertion of the correspondence between colour and refraction. This concept was not only confirmed and developed in the demonstration of the theorem but it led also to a *Corollarium*:

> It follows that the rays which carry the same colour are parallel.[31]

Here we find once again the mediaeval concept that colour was like a *species* propagated along rays; but Marci attributed the same colour to the rays which emerged from a prism parallel to each other. There is no doubt that Marci had definite ideas concerning refraction and dispersion through a triangular prism, even though he was unable to explain the phenomenon. Marci appeared to have reached a much greater accuracy in his measurements than either his predecessors or his successors for many years after. This was because all the others carried out experiments using the *lumen* from the Sun, which although a point source had a diameter of half a degree. Marci on the other hand, in his experiments, used a candle as a source of light. This was much smaller than the Sun, and allowed him to carry out much more accurate observations. Today we use a narrow slit for the study of spectra produced by a prism.

Also of great interest are Theorems XIX and XX. The first stated that:

> A coloured ray which undergoes a second reflection does not change its depth of colour.

and the second:

> A coloured ray which undergoes a further refraction does not change colour.

As if all this was not enough, Marci in his Theorems XXI and XXII, insisted on the correspondence between colour and deviation, following refraction. The first of these theorems stated:

[31] Op. Cit. p. 100

> Similar but unequal light produces equal colours by means of the same refraction.

and the second:

> Dissimilar but equal light produces unequal colours by the same refraction.

In the course of the demonstration of these theorems, Marci himself explained the meaning he attributed to the terms he used. Thus he called 'similar' the beams of rays which reach the prism with equal angles of incidence, and he called 'unequal' beams which did not have the same intensity. This particular care which Marci took in making the colour independent of the intensity of the beam of light, represented his attempt to free himself from the mediaeval ideas, according to which the 'apparent' colours produced by refraction were related to the luminous intensity of the beam. Yellow and green were to be seen where the beam was more intense, while red and violet represented a transition towards the black of darkness. In this way Marci linked explicitly the colour with the deviation produced by refraction by the prism, and freed it from the intensity of the beam. There is no doubt that Marci introduced ideas which we may call definitive even if based on premises that very soon were to be completely demolished and replaced by new ideas.

As in the case of Kepler, Marci's work was not well received by anyone and was soon forgotten. We shall soon see that his conclusions later came to the fore with stronger supporting evidence. The contribution of Marci have recently come to be known, simply because his work has been very carefully studied on the occasion of the tercentenary of his death.

Let us now proceed to the examination of the work of the two great men, Isaac Newton and Christian Huygens, whose work and fame overshadowed all the work of their predecessors and of their contemporaries.

CHAPTER FIVE

Newton and Huygens

ALTHOUGH Christian Huygens (1629–1695) was thirteen years older than Isaac Newton (1642–1726), we shall give precedence to Newton because Huygens became interested in the theory of the nature of light only at an advanced age. He had devoted most of his time, with great success, to questions of geometrical optics and optical instruments, subjects with which we are not directly concerned; his *Traité de la lumière* was not written until 1678. Newton, on the other hand almost from the beginning of his work, devoted himself to what was one of the most challenging questions of the century. His early works from 1666 onwards dealt with the nature and properties of colours.

Newton himself, in a letter to Oldenburg, then secretary of the Royal Society, dated February 6th, 1671–2, related how it was that he began his studies of colours and hence of optics. This letter begins thus:

> Sir,
>
> To perform my late promise to you, I shall without further ceremony acquaint you, that in the beginning of the year 1666 (at which time I applied myself to the grinding of Optick glasses of other figures than *Spherical*) I procured me a Triangular glass-Prisme, to try therewith the celebrated *Phaenomena of colours.* And in order thereto having darkened my chamber, and made a small hole in my window shuts, to let in a convenient quantity of the Sun light, I placed my Prisme at its entrance, that it might be thereby refracted to the opposite wall. It was at first a very pleasing divertisement, to view the vivid and intense colours produced thereby; but after a while applying myself to consider them more circumspectly, I became surprised to see them in an *oblong* form; which according to the received law of Refraction, I expected should have been circular. . . .

The letter continues with a detailed description of all the reasoning which the young Newton used to try to find an explanation for this 'oblong form' which he expected to be circular. At first he attributed this to some faults in the glass and he carried out various tests which led him to reject this supposition. He then considered the possibility that the rays, after having gone through the prism, could follow a curved path (he quoted as an example a tennis ball which is struck obliquely by a racquet), and he carried out experiments which enabled him to discard this theory. After all this Newton performed another experiment which he himself called the '*Experimentum Crucis*'. He let the beam of rays which emerged from a prism, fall upon a screen which had several holes, so that from any of these holes a narrower beam emerged. He caused one of these beams to fall upon another prism which had its sides parallel to those of the first prism and as a result of this he noticed that the 'oblong form' was not present any longer. From this he concluded that the oblateness of the coloured image originally observed was due to the constitution of the original beam falling on the prism, and that the beam had to consist of several rays each one being 'differently refrangible'. Still in the above mentioned letter he stated:

> . . . And so the true cause of the length of that Image was detected to be no other, than that *Light* consists of *Rays differently refrangible*, which, without any respect to a difference in their incidence, were, according to their degrees of refrangibility, transmitted towards divers parts of the wall. . . .

The letter continues:

> When I understood this, I left off my aforesaid Glass-works; for I saw, that the perfection of Telescopes was hitherto limited, not so much for want of glasses truly figured according the prescriptions of Optick Authors, (which all men have hitherto imagined) as because that Light itself is a *Heterogeneous mixture of differently refrangible Rays*. . . .

As a result of this he thought of making telescopes which had for objective a parabolic mirror instead of a lens. Indeed he did make such telescopes and carried out observations of Jupiter's satellites and of the phases of Venus. He also conceived the idea of making a reflecting microscope.[1] Perhaps we should note here that the young Newton – he was then in his early twenties – began making

[1] *The correspondence of Isaac Newton* Published by the Royal Society 1959, Vol. I, p. 92 and 95

telescopes just like present-day amateur astronomers; his interest was purely practical, so much so that as soon as he realized that light was composed of 'rays differently refrangible' he avoided the complications produced by this, by using reflectors instead of refractors. Newton's discovery of the oblateness of the refracted beam of light was mainly due to the smallness of the light source, in this case the small hole he had made in the shutter of the window of his room (as Grimaldi himself had done earlier). So once again the use of a small source led to important results, yet this method was still rarely used.

These experiments by Newton began in the year following the publication of *De Lumine* by Grimaldi. There is little doubt, as we shall see later, that Newton was acquainted with this work. Probably the mention of the 'celebrated phaenomena of colours' in the letter quoted above, was a reference to the investigation that Grimaldi made, and recorded in his book, into the behaviour of prisms and the nature of colours. After some time involved in practical work, Newton finally realized the importance that theoretical as well as experimental studies could have in the field of colour. He abandoned telescopes and microscopes to devote himself to the development of experiments of a more scientific nature. It was probably at this time that the fundamental idea, which was to lead him for the rest of his life, took shape in his mind. He found a path that he never attempted or, indeed, wished to leave, notwithstanding the seemingly insurmountable obstacles which he met, which were a source of great vexation for him. As we have seen from the work of Grimaldi, refraction at that time was still a mystery. It was not enough for the new scientific method to describe natural phenomena in the same way as in the past; what was demanded now was an explanation. The new fashion was to produce a model to explain phenomena, but alas, in the case of refraction no one seemed to be capable of producing such a model.

Newton, in the letter quoted above, suggested: '... if the rays of light should possibly be globular bodies ...' they would be attracted by material bodies. If the incidence was oblique they would be deviated from the trajectory of incidence. Furthermore, if the globular bodies had different masses, they would also be subject to different deviations or, in other words, they would be 'differently refrangible'. This seemed to be the explanation of the observed phenomena. It was obvious: the incident beam of light which appeared white, consisted of corpuscles of different masses; refraction by a prism had the effect of dividing them, selecting them and sending them in different directions according to their mass, and their colour; therefore colour and refrangibility were two phenomena linked together. The picture was indeed attractive, and we should not be surprised to find that the young Newton was fascinated by it and decided to make every possible effort to obtain an experimental proof which would be irrefutable.

Newton's letter, which gave the first description of his experiments and of his ideas, was read to the Royal Society on February 8th, 1671–2 and was published in the *Philosophical Transactions* of 1672, Volume VII (p. 3075). The history of the experiments and the theory which was to be called after Newton, is very complex and very important. We shall have to discuss it in detail because the theory had repercussions which are still continuing. If we wish to make a general appreciation of its significance and of its value as objectively as possible, we must study the work which contains these ideas, namely *Opticks*, which Newton published about forty years later in 1704. The fact that this work was published at such a late date leads us to believe that it contains the complete studies and the fully developed ideas of Newton, particularly since the various editions which followed do not show any substantial changes. On the other hand, *Opticks* was first published forty years after Grimaldi's *De Lumine*, and nearly twenty years after Huygens' *Traité de la lumière*, therefore it is reasonable to expect to find in it a general outline of Newton's important contribution to the progress of optics.

Opticks is divided into three Books. Book I is subdivided into two parts and deals with refraction, dispersion, and the analysis and the synthesis of colours. Book II is divided into four parts and contains the study and the interpretation of the phenomena which nowadays we would call interference. Book III deals with diffraction. In Book I we also find axioms and definitions, while Book III contains the conclusions reached. (Fig. 62.)

There is no doubt that *Opticks* is an extremely important and interesting work which deserves a very detailed study. But if we want to follow without bias the evolution of the ideas concerning the nature of light in the seventeenth and eighteenth centuries, it will be necessary to call attention not only to its merits, which indeed are very great, but also to its weaknesses, on which we shall have to dwell because they represent a case of exceptional interest for the history of light.

As it has been stated in the case of other authors, to understand completely what Newton wrote in the first pages of his work it is necessary to have read and studied the book right to the end. He began his work thus:

> My design in this book is not to explain the properties of light by hypotheses, but to propose and prove them by reason and experiments: in order to which I shall premise the following definitions and axioms.[2]

It would seem that Newton wished to prove experimentally the Propositions which he gave in his work without using any premises or preconceived theories. Often under the title of some of the Propositions we find 'The Proof by Experiments'. A similar approach is often found in Grimaldi's *De Lumine*. It was the fashion in the seventeenth century to follow this method as a reaction to the classic method of the Peripatetics. We cannot deny that as an approach it was excellent, but we cannot deny either that in reality it was only an illusion. The great value of theories is to act as guide to experiments, indeed it is not possible to carry out systematic and conclusive experiments without some theoretical guidance, be it explicit or not. The critical examination of the 'Definitions' shows that on the whole Newton acted like all research workers.

The first 'Definition' is very important, it stated:

> By the rays of light I understand its least parts, and those as well successive in the same lines, as contemporary in several lines.[3]

[2] Isaac Newton, *Optics* Great Books of the Western World, Edited by Robert Maynard Hutchins, Encyclopaedia Britannica Inc. William Benton, 1952 Book I, Part I p. 379

[3] Op. Cit. p. 379

There is no doubt that this was a new idea, a new concept of the luminous ray, but we must ask ourselves whether the corpuscular theory of light was not already included here. If it was not, then what did this definition mean? There followed an explanation so interesting that it deserves to be given in full.

> For it is manifest that light consists of parts, both successive and contemporary; because in the same place you may stop that which comes one moment, and let pass that which comes presently after; and in the same time you may stop it in any one place, and let it pass in any other. For that part of light which is stopped cannot be the same with that which is let pass. The least light or part of light, which may be stopped alone without the rest of the light, or propagated alone, or do or suffer any thing alone, which the rest of the light doth not or suffers not, I call a *ray* of light.

Today this would be called a '*quantum*'.

We must note here that for Newton the ray was not the trajectory of light but was that 'least light' which could be isolated and was independent of its trajectory. Following this idea he introduced expressions which we still use conventionally nowadays, but that he used literally. For example, when in some books we read that the rays *pass* through a point, literally this implies that the rays are moving, but what is meant is that 'something' moves along the trajectories called rays. This is also established in Definition II which concerns the concept of refrangibility:

> Refrangibility of the rays of light, is their disposition to be refracted or turned out of their way in passing out of one transparent body or medium into another. And a greater or less refrangibility of rays is their disposition to be turned more or less out of their way in like incidences on the same medium.[4]

At this point the unbiased student is bound to ask himself what Newton wished to convey with this definition. It must be admitted that he introduced new ideas inasmuch as he postulated that refrangibility and its variation (namely dispersion) is a 'disposition'

[4] Op. Cit. p. 379

of the rays and not of the media. This would be acceptable if the explanation which followed did not confuse us on what Newton was really thinking.

> Mathematicians usually consider the rays of light to be lines reaching from the luminous body to the body illuminated, and the refraction of those rays to be the bending or breaking of those lines in their passing out of one medium into another. And thus may rays and refractions be considered, if light be propagated in an instant. But by an argument taken from the equations of the times of the eclipses of Jupiter's satellites, it seems that light is propagated in time, spending in its passage from the Sun to us about seven minutes of time: and, therefore, I have chosen to define rays and refractions in such general terms as may agree to light in both cases.[5]

We can conclude from this that Newton had good intentions, but we must recognize that he had undertaken a superhuman task.

We come now to Definition III:

> Reflexibility of rays is their disposition to be reflected or turned back into the same medium from any other medium upon whose surface they fall. And rays are more or less reflexible which are turned back more or less easily.[6]

Why postulate reflexibility and the variation of reflexibility as an attribute of the rays rather than of the reflecting body? This definition contains extremely complicated ideas which can only be understood when the greater part of Newton's work has been carefully studied.

In the Definitions which follow, Newton examined angles of incidence, reflection, refraction, their sines and the meaning of expressions such as homogeneous light, primary, homogeneous and simple colours. There follow eight 'Axioms', the first five of which refer to the laws of reflection and of refraction in the form

[5] Op. Cit. p. 379. This is an obvious allusion to Roemer's measurements. Note the rather cautious noncommittal form adopted by Newton. This could be taken as a symptom that Roemer's ideas were not fully accepted.

[6] Op. Cit. p. 379

expressed by Euclid and by Descartes. The other three axioms summarize the basic ideas of elementary geometrical optics, namely the formation of real and virtual images. In order not to misunderstand the meaning of the word axiom let us quote the words of the author himself. 'I have now given in Axioms and their explications the sum of what hath hitherto been treated of in Optics . . .'[7] Newton appeared to accept these axioms as fundamental principles and on this basis he began his construction.

The experimental part of Newton's work is truly magnificent. It is carried out with great accuracy, method and with rare ability. Every proof is accompanied by a counter-proof, and reveals in the young physicist the great mind of the scientist. Unfortunately he was not always an impartial and objective observer; sometimes we have the impression that the observer was a man determined to find a confirmation of his own theories and views.

Newton again took up the experiments on refraction through prisms that had been made by Descartes, Marcus Marci, Grimaldi and others, but he carried them out with great skill and obtained outstanding results. He still made use of a beam of light from the Sun, that is a beam having an angular width of 32 minutes of arc, but by carrying out his observations at distances of several yards from the prism and by using prisms made of highly dispersive substances, he was able to make very accurate measurements from which he deduced important new explanations. Particularly valuable was his experiment with two prisms at right angles (in a 'cross position') which together with other experiments led him to assert the existence of the relationship between 'refrangibility and the colour of light'. Newton not only confirmed that the ratio between the sine of the angle of incidence and the sine of the angle of refraction was constant, but went further and asserted that those of his predecessors who had made similar experiments (figs. 64 and 65)

> . . . do acquaint us that they have found it accurate. But whilst they, not understanding the different refrangibility of several rays, conceived them all to be refracted according to one and the same proportion, 'tis to be

[7] Op. Cit. p. 386

> presumed that they adapted their measures only to the middle of the refracted light. . . .[8]

He, on the other hand, was able to show that this ratio was different for different colours, and in spite of very difficult conditions, he succeeded in making accurate measurements. Thus the nature of colour began to acquire a different meaning for Newton, almost a tangible and experimental reality which must have been as exciting for Newton himself as for his contemporaries.

These assertions and some of the experiments later described, indicate that Newton and also the members of the Royal Society to whom his discoveries had been communicated, were not aware of the *Thaumantia Liber* by Marci, to which we have referred in previous pages (see p. 156). That Newton should be unaware of the work of Marci is, in a way, understandable, because throughout the ages young men are inclined to ignore the work of their immediate predecessors, relying more for their knowledge on classic texts. We shall see more examples of this later on. What seems to be less understandable is that Marci's work should have been unknown to eminent members of the Royal Society. It is remarkable that no one, not only among English scientists, claimed the priority of Marci in connection with several ideas that contributed so much to establish the fame of the young English scientist.

This is an interesting question which has recently been investigated. It appears that Marci had been in touch with the Royal Society and that the Royal Society had asked one of their members who was travelling on the Continent at the time, to visit Marci in Prague. Unfortunately when this messenger reached Prague, Marci had died. The fact that Marci's ideas were not mentioned either by Newton or by other scientists of the time indicates that the *Thaumantia Liber* had been completely forgotten. Perhaps the explanation of this lies in the difficulties of communication and of cultural exchange between various parts of Europe because of the Thirty Years War. Another reason may be that Marci's book followed a traditional line, and in the seventeenth century,

[8] Op. Cit. p. 409

when ideas developed and evolved with a speed almost comparable to that of the present day, books tended to become out of date very quickly and soon lost most of their interest and appeal. More important books than that of Marci underwent the same fate, those of Kepler, for example, and of Grimaldi not much later. There is little doubt that this process was accelerated by the renown that Newton quickly acquired throughout the scientific world. Inevitably scholars tended to concentrate on his works and to neglect the work of other writers.

The rediscovery of Marci's *Thaumantia* today, may at first suggest some doubts as to whom the credit should be given for certain discoveries. Newton may not seem to merit the fame of being the first to have determined the relationship between 'colour and refrangibility' or to have achieved the experimental proof that monochromatic light did not undergo further dispersion when refracted by successive prisms. Nevertheless we must recognize that, while Marci described these phenomena using mediaeval theories concerning colour, it is incontestable that Newton, by means of his scientific theory, explained refraction as an effect of material attraction between a refracting material body and the material corpuscles of light. Even if this theory did not last for long, it served its purpose in the general evolution of ideas in this particular field, by giving a mechanistic explanation of the phenomena of light, and this explanation was of invaluable philosophical importance in that particular phase of the evolution of scientific ideas in general.

Newton carried out some very important experiments. He showed that, in general, light consisted of several components; with a prism he succeeded in separating the individual components, one of which he isolated by means of diaphragms. He allowed it to fall upon another prism and discovered that this component could not be divided further. This was the beginning of the analysis of light which later evolved into modern spectroscopy. From this analysis Newton introduced another example of a division into elementary units, similar to that adopted by Alhazen to explain the mechanism of vision, and that used by Grimaldi to discover diffraction. These two scholars had divided the geometrical dimensions of the beam of light into elements; Newton did the same for the internal structure of light. Today modern optics shows that the two processes are comparable in most respects. After analysing light or, as we would say today, after obtaining monochromatic light, Newton recombined the components and obtained white light again, a white light which like the original light could be divided into monochromatic light. He showed that the statement of Descartes, which said that refraction gave colour to light only in the region of separation between light and shadow, was false. He gave his conclusions in Proposition II of the Second Part of Book I.

> All homogeneal light has its proper colour answering to its degree of refrangibility, and that colour cannot be changed by reflexions and refractions.[9]

From this he proceeded to give the Definition of the concept of colour in a precise and unmistakable manner:

> The homogeneal light and rays which appear red, or rather make objects appear so, I call rubrific or red-making; those which make objects appear yellow, green, blue and violet, I call yellow-making, green-making, blue-making, violet-making, and so of the rest. . . .[10]

Newton stressed that when he talked of seeing coloured rays or coloured light, he was only expressing himself in a manner of speaking, not 'philosophically and properly'. These rays were not coloured but: 'In them there is nothing else than a certain power and disposition to stir up the sensation of this or that colour. For as sound in a bell or musical string . . . is nothing but a trembling motion' which reaches our ear. Having thus defined colour, Newton classified seven colours, presumably only to continue the tradition which in earlier ages had chosen this number. He excluded white and black, and put the others in the following order: red, orange, yellow, green, blue, indigo and violet. Although he was fascinated by all this, yet he did not dare to go too far, and he added some reservations:

[9] Op. Cit. p. 427
[10] Op. Cit. p. 428

> I speak here of colours so far as they arise from light. For they appear sometimes by other causes, as when by the power of phantasy we see colours in a dream, or a madman sees things before him which are not there; or when we see fire by striking the eye, or see colours like the eye of a peacock's feather by pressing our eyes in either corner whilst we look the other way.

Faced with the situation that colours could exist even without light, he appeared unconcerned and simply glossed over the whole question by adding:

> Where these and such like causes interpose not, the colour always answers to the sort or sorts of the rays whereof the light consists....

Glossing over difficulties is one way to deal with complex problems both in science and in philosophy. It is the tactic followed by those who accept a given idea and make it an act of faith. Those who are not prepared to do this are condemned to doubt and to dissatisfaction, and we have a typical example of this in Book II of *De Lumine* by Grimaldi. For a great thinker the glossing over of difficulties is a very dangerous practice. His disciples tend to follow him blindly and it is impossible to foresee for how long objections will remain undetected. The day such objections are discovered is the day of reckoning when the whole edifice patiently built up collapses.

Although Newton stated that he would describe experimental results independently of any hypothesis on the nature of light, almost in every sentence we have a glimpse of the idea which was taking shape in his mind, namely that light consisted of a swarm of particles moving through space at a great speed and each one having a different mass according to its colour. The quotation given above that '... the colour always answers to the sort or sorts of the rays whereof the light consists ...' betrays the thoughts of the writer, and we remember that this is not a question of geometrical rays as defined by 'mathematicians', but of 'parts both successive and contemporary'. What Newton was saying was that light was composed of different types of these rays. Probably he became increasingly convinced of this when he realized the possibility of explaining refraction by the mass of the particles, and of explaining the difference in refraction, or dispersion as we would say today, by the variety of mass of the particles. This question is discussed in Section XIV of *Philosophiae naturalis principia mathematica* under the title 'The motion of very small bodies when agitated by centripetal forces tending to the several parts of any very great body'. Here the author showed that once we accept the existence of attraction between the corpuscle and the large body limited by a plane surface, which is naturally transparent for the corpuscle, the latter will penetrate the larger body, and in doing so will deviate from the rectilinear trajectory which it was following. The corpuscle, after penetrating the larger body, will follow a new rectilinear trajectory connected to the previous one by a very small arc of a curve; the ratio between the sine of the angle of incidence and the sine of the angle of refraction will be the same as the ratio between the velocities of the corpuscle before and after crossing the surface of the larger body. This was the reasoning which Descartes had expressed formally, and which Newton explained by means of the force of attraction of the denser body acting upon the corpuscle which constituted light. This attraction increased the component of the velocity normal to the surface of separation of the media, thereby compelling the trajectory to bend towards the normal to the surface in the denser medium. This required that the resulting velocity of the corpuscle should be greater in this denser medium. Descartes had justified, or at least he thought he had justified, this conclusion; Grimaldi thought that this conclusion was absurd 'in itself and in its sole exposition'; Newton did not dwell on it.

In the question of dispersion we must mention another case which has become famous. In discussing the 15th Experiment of Book I, Part One of *Opticks*, that is the experiment concerning two prisms which were crossed, Newton remarked that the resulting spectrum, as we call it today, was rectilinear. As we have already mentioned, Newton experimented with a whole beam of sunlight and therefore could not detect the very small bending that the spectrum undergoes when prisms of different types of glass or different substances are used. This bending is only detectable with modern instruments of high precision. Newton, however,

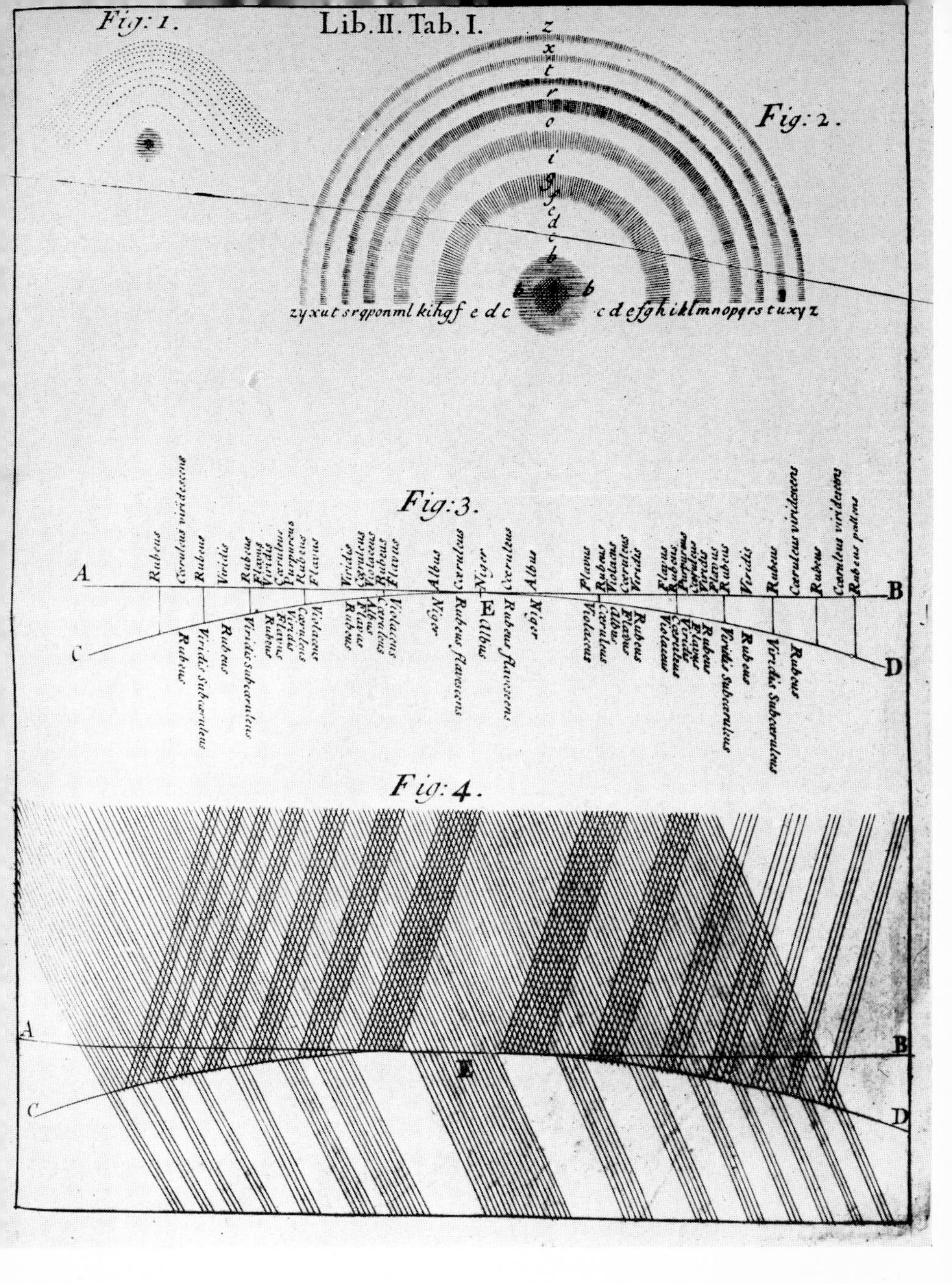

FIG. 67. Diagrams showing the "rings", from *Opticks*

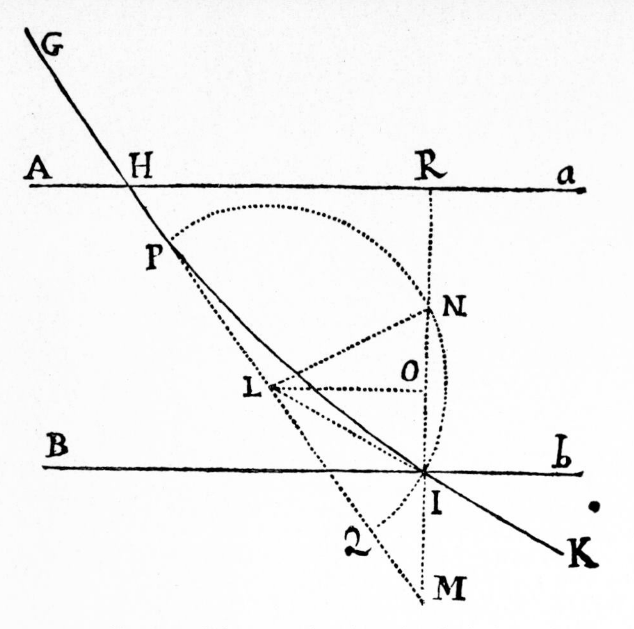

FIG. 68. Diagram showing the deviation of a light ray when passing from one medium to another, from *Philosophiae naturalis principia mathematica*

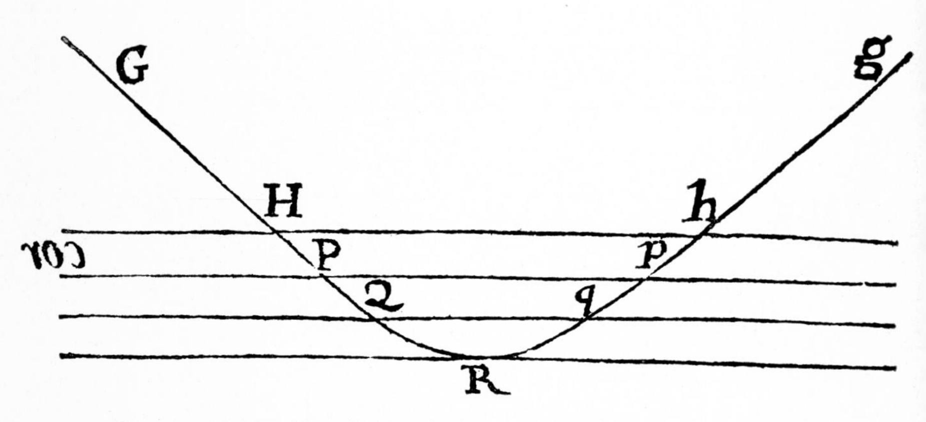

FIG. 69. Mechanism of the reflection of light rays according to Newton, from *The Principia*

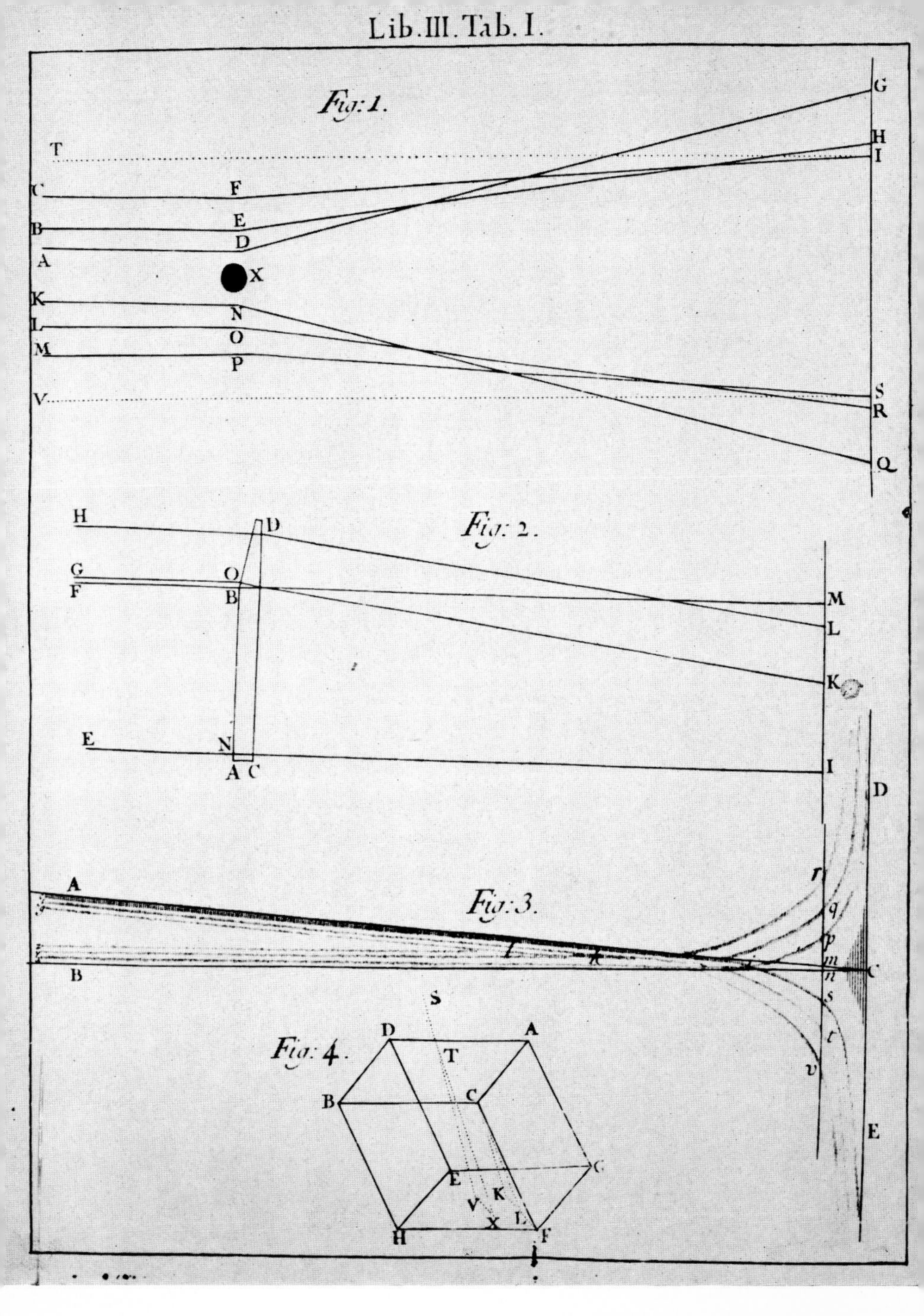

FIG. 70. Diagram showing experiments of diffraction from *Opticks*

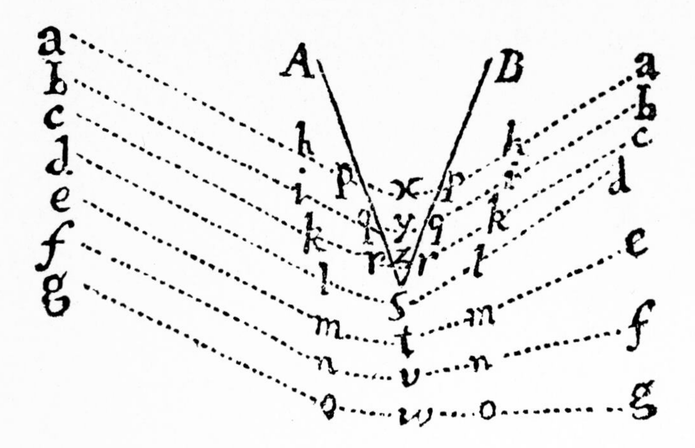

FIG. 71. Mechanism of diffraction according to Newton, from *The Principia*

FIG. 72. Diagram showing the propagation of waves through a hole in a screen, from *The Principia*

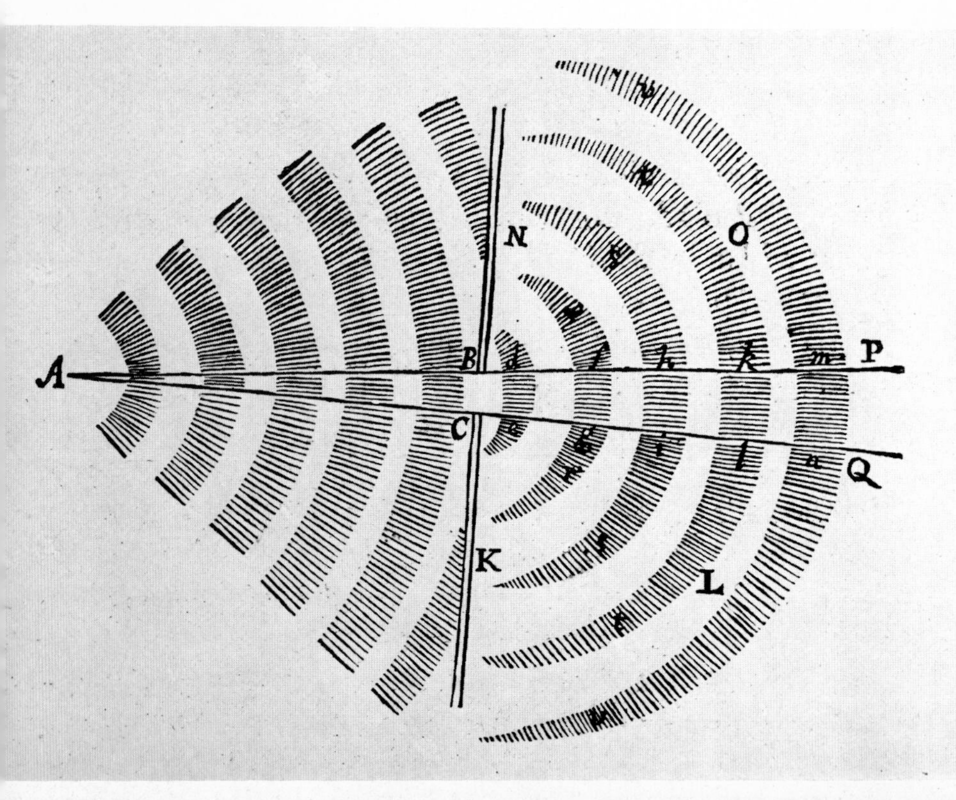

showing a remarkable experimental skill, noticed that something unusual was taking place, and his intuition led him to a geometrical explanation of this experimental anomaly. Strangely enough, having gone so far he changed his method: he abandoned his original inductive-experimental reasoning, and instead deduced what the behaviour of prisms would have been according to the hypothesis: 'that bodies refract light by acting upon its rays in lines perpendicular to their surfaces'. Newton was so convinced that this hypothesis explained satisfactorily his theory of the 'oblique' spectrum that later, following a geometrical demonstration, he concluded:

> And this demonstration being general, without determining what light is, or by what kind of force it is refracted, or assuming anything further than that the refracting body acts upon the rays in lines perpendicular to its surface; I take it to be a very convincing argument of the full truth of this Proposition.

This certitude caused him to desist from trying to extend his experiments, and led him to expound a rule which has come to be known in history of science as 'Newton's error'. This is particularly important in the history of light, because this error was necessary to the existence of Newton's theory on the structure of light. In fact, experimenting with only a few substances, mainly glass or water, and with a few glass prisms, he mentioned a greenish one and a clear one, he concluded that the dispersion was proportional to refraction.[11] This conclusion became famous as 'Newton's error', because it denied the possibility of making achromatic lenses and prisms. Newton himself, because he was convinced that it was impossible to eliminate the chromatic aberration of the lenses of a telescope, was led to study the use of concave metal mirrors in telescopes.[12]

[11] Op. Cit. Book I, Pt. I, Experiment 3, p. 390

[12] Op. Cit. p. 420

It may be rightly said that this is a true technical aspect of the question, while we are more interested in the theoretical side. If it was accepted that light was composed of material particles having a different mass according to colour, then their deviation should have been a function of the mass of the particles themselves, with the mass of the attracting body intervening in each case as a fixed factor. Thus the constancy of the ratio between refraction and dispersion was necessary for the corpuscular theory of light. Experiments would have shown that this was not true, but Newton did not do these experiments. This is particularly strange when we consider that he carried out extremely delicate and difficult experiments with the means he had available, in order to determine the value of the ratio between refraction and dispersion (Proposition VII). Newton obtained for one case the value of 27.5 and he concluded that this would be the value for every case and that, therefore, in refracting telescopes the chromatic aberration would always be greater than the spherical aberration. The only way out was to use reflecting telescopes. In conclusion, this first part of Newton's work can be considered as a great and definite experimental progress. Colours were definitely detached from the old-fashioned philosophy, were made objective, defined, measured and analysed; in addition a theory was put forward, based on the model of corpuscular light and of gravitation, which attributed a different mass to different colours. The progress would have been wonderful had it not left behind two very dangerous questions, namely the greater velocity of light in denser media and the relationship between refraction and dispersion. But an even greater doubt endangered the theory: what were the colours which were not produced by light? Newton's troubles did not end here. More dangerous threats against the corpuscular hypothesis were encountered when he tried to explain the colours of thin films and diffraction.

In Book II of *Opticks*, Newton returned to Hooke's experiments on the colour of thin films and plates, and produced another masterly series of experiments with measurements of very high precision and of special interest. He had two particularly ingenious ideas; the first, which was a direct consequence of previous research, was to carry out experiments both in white light and in monochromatic light produced by a prism. The second was to use the spherical form of the surface of contact to measure the thickness of the very tenuous layer in which colours were produced. Hooke had already stated that the thickness of the layer was the fundamental element which produced the various colours, but he had assumed this as an hypothesis, since he had been unable to detect the presence of this thin layer by any direct means. While Hooke tried to measure the thickness of the layer directly with a microscope, Newton followed an indirect method. He placed a spherical and a plane surface against each other, and from the radius of the spherical surface he was able to deduce the form of the thin layer of air between the two. In these conditions, as we know, a series of concentric rings are formed around the point of contact of the two surfaces. When we observe the reflection from the side of the plate where the light falls, all luminosity disappears, whether we use white or monochromatic light. The radii of the rings increase in the same ratio as the square root of the number of their order, and the radii of the rings, occupying a given position in the series, increase as the colour of the light changes from violet to red.

Newton discovered all these laws. He was struck by the fact that while in white light he could see only eight or nine rings, in monochromatic light he could see several dozens; but he was even more surprised when he looked through a prism at the rings produced by white light. Without the prism he could see the usual eight or nine rings, but through the prism there appeared different systems of rings of various colours and displaced with respect to one another. Each system appeared to consist of many rings, as if they had been obtained by monochromatic light (fig. 67)

> . . . insomuch that I could number more than forty, besides many others, that were so very small and close together that I could not keep my eye steady on them severally so as to number them, but by their extent I have sometimes, estimated them to be more than a hundred.[13]

This extraordinary experiment suggested to Newton that in a thin film when white light was used all the various systems of rings of different colours were present independent of one another.

[13] Op. Cit. Book II, Part I, Observation 24, p. 469

Because of their different dimensions, the outermost rings became indistinct and only the broader rings, which were nearer to the centre and were common to all the rings, became more easily distinguishable. This was a conclusion which agreed very well with the model of light obtained by the experiments with prisms. The great difficulty, however, was to explain the formation of rings in monochromatic light. Newton's measurements and calculations established that the dark rings were always formed where the thickness of the film was a multiple of a given value, namely 1/89,000 part of an inch (0.285μ) for light of a bright yellow colour, and that the bright rings occurred where the thickness had a value which was between those of the dark ring immediately preceding and following. In addition Newton also noted that the ratio of the thicknesses required to produce red and violet rings was 14 to 9. These experimental results showed an obvious periodicity in the behaviour of light, a periodicity which was characteristic of each colour! There was something else remarkable. Rings could be seen not only by looking at the film or plate from the side of the incident light (by reflection), but also by looking at it from the opposite side (by transparency), and these rings were complementary to the others in the sense that the bright centre and the bright rings occupied the positions of the dark rings when observed by reflection. Newton represented this phenomenon clearly in a drawing in his book. He linked this phenomenon with that which had given so much trouble to Grimaldi, namely the reflection from both sides of a transparent slab of glass; he also linked the phenomenon with the reflection by coloured bodies as well as with refraction. The result was a picture more attractive than convincing, which appeared wonderful to Newton's admirers, but not entirely satisfactory to Newton himself. The framework of the theory is contained in a sentence in Definition III which is barely noticeable and to the superficial reader may appear purely as an example of simple pedantry:

> . . . and rays are more or less reflexible which are turned back more or less easily.[14]

[14] Op. Cit. Definition III, p. 379

Let us remember Newton's definition of a ray: 'the least light or part of light'. If we wish to follow more easily the model on which he based his terminology we must think of a corpuscle. With this in mind, the sentence of Definition III, quoted above, cannot be interpreted as what would today be called 'the coefficient of reflection' of various substances; it represents a completely different concept. First because it is the rays which are more or less reflexible, and this selective property is postulated for the rays without any reference whatsoever to the reflecting body. Secondly because it is not a question of a statistical effect, in the sense that in reflection from a surface, a given percentage of the corpuscles enter the body or avoid it by rebounding. As the writer stated clearly, it was a question of a property or 'disposition' because of which certain rays (corpuscles) were better reflected than others, in external conditions which were identical. The less well reflected rays were those which penetrated better inside the body that was limited by the reflecting surface. This 'disposition' did not agree exactly with the mechanism of reflection that Newton had decisively supported earlier. According to him, matter should be considered as being composed of wide gaps and a few material particles. Here he seemed to have adopted the second model of Descartes, but to have given it a new lease of life. He proved the 'porosity' of matter with arguments which were not more detailed than those used by Grimaldi. In a matter so rarefied, 'the cause of reflexion is not the impinging of light on the solid or impervious parts of bodies as is commonly believed'.[15] The reasons given by Newton are as follows:

> First, that in the passage of light out of glass into air, there is a reflexion as strong as in its passage out of air into glass, or rather a little stronger, and by many degrees stronger than in its passage out of glass into water....

Grimaldi had tried in vain to solve this mystery.

> Secondly, if light in its passage out of glass into air be incident more obliquely than at an angle of 40° or 41° it is wholly reflected [total

[15] Op. Cit. Book II, Part III, Proposition 8, p. 485

> reflection] if less obliquely [in modern terms critical angle] it is in great measure transmitted. . . . Thirdly, if the colours made by a prism . . . be successively cast on a second prism . . . the second prism may be so inclined to the incident rays that those which are of a blue colour shall be all reflected by it, and yet those of a red colour pretty copiously transmitted. . . .
>
> Fourthly, where two glasses touch one another [in modern terms, optical contact], there is no sensible reflexion. . . .
>
> Fifthly, when the top of a water bubble . . . grew very thin there was such a little and almost insensible quantity of light reflected from it that it appeared intensely black; whereas round about that black spot . . . the reflexion was so strong. . . . Sixthly, if reflexion were caused by the parts of reflecting bodies, it would be impossible for thin plates or bubbles, at one and the same place, to reflect the rays of one colour and transmit those of another. . . .
>
> Lastly, were the rays of light reflected by impinging on the solid parts of bodies, their reflexions from polished bodies could not be so regular as they are. . . . So, then, it remains a problem how glass polished by fretting substances can reflect light so regularly as it does. . . .

In all this there is enough evidence to exclude the possibility of reflection by an elastic collision of the luminous corpuscles with matter. The mechanism which Newton substituted for the one he had demolished is described in Proposition XCVI, Theorem L of *Principia*, and is also given in Proposition 10 of *Opticks*.

> If light be swifter in bodies than in vacuo, in the proportion of the sines which measure the refraction of the bodies the forces of the bodies to reflect and refract light are very nearly proportional to the densities of the same bodies; excepting that unctuous and sulphureous bodies refract more than others of this same density.[16]

So this is the explanation: bodies attract the corpuscular luminous rays. In certain conditions these rays penetrate the body and are bent towards or away from the normal to the surface of separation, according to whether their velocity in the body is respectively greater or smaller than in the outside medium. In other conditions we have a bending of the trajectory, as in a mirage, and reflection

[16] Op. Cit. p. 488

takes place. If all the particles are in such mechanical conditions that they are reflected, then we have total reflection. In its mechanical simplicity this explanation is very attractive, but why should fatty and sulphureous bodies have a different behaviour? This objection was very serious for Newton who was trying to prove that reflection and refraction were among the phenomena due to universal gravitation. He tried to overcome this difficulty by stating:

> . . . And as light congregated by a burning glass acts most upon sulphureous bodies, to turn them into fire and flame, so, since all action is mutual, sulphurs ought to act most upon light. . . .[17]

To prove this reciprocity of action between light and matter, Newton added: '. . . that the densest bodies which refract and reflect light most strongly grow hottest in the summer Sun, by the action of the refracted or reflected light . . .', but he did not give any example. The text gives the impression that the author did not wish to discuss this question in detail. An even more serious objection existed. If the effect of the body on the luminous particle was excluded why was it that sometimes this luminous particle was reflected and at other times it was refracted, although the mechanical conditions were identical? There was only one possible solution. Since the conditions of the environment were identical for the two particles, one of which was reflected and the other refracted, it was not possible to explain the phenomenon by having recourse to any external action. It remained, therefore, only to postulate a 'disposition' of the particle itself, so that those wishing to pass did so, those not wishing to pass did not. If this was the case, what was then the value of the whole mechanical theory?

Newton showed great skill in presenting all this as a great victory. Referring to the experiments and to the measurements of the rings, he presented this 'disposition' as if it were a law, a discovery which he expressed in Proposition XII.

[17] Op. Cit. p. 491

> Every ray of light in its passage through any refracting surface is put into a certain transient constitution or state, which in the progress of the ray returns at equal intervals, and disposes the ray at every return to be easily transmitted through the next refracting surface, and between the returns to be easily reflected by it.[18]

From this he could explain the rings, and he reached the following definition:

> The returns of the disposition of any ray to be reflected I will call its fits of easy reflexion, and those of its disposition to be transmitted its fits of easy transmission, and the space it passes between every return and the next return, the interval of its fits.[19]

Unfortunately Newton no longer appeared to be master of the situation. He postulated this 'disposition' of the corpuscles to be reflected or to penetrate better. He also postulated the periodical variation at regular intervals with the pretext that this explained satisfactorily the formation of rings. He made this transitory 'disposition' arise at the moment of refraction through the surface of entry of a thin plate, and then made it disappear after it had passed through the surface of emergence. A few pages further on he expanded the field of action of this 'disposition' to explain what appeared to be a very simple phenomenon, which was in fact very complex, namely the partial reflection upon the two sides of any transparent plate. Hence in Proposition XIII he stated:

> The reason why the surfaces of all thick transparent bodies reflect part of the light incident on them, and refract the rest, is that some rays at their incidence are in fits of easy reflexion, and others in fits of easy transmission.[20]

He added:

> ... And probably it (light) is put into such fits as its first emission from luminous bodies, and continues in them during all its progress. ...

[18] Op. Cit. Book II, Part 3, p. 492

[19] Op. Cit. Book II, Part 3, p. 493

[20] Op. Cit. Book II, Part 3, Proposition XIII, p. 493

In all these pages we feel that the author himself is not really satisfied with these ideas. The feeling is that, had he been truthful, he would have limited himself to stating: that the rings suggested a periodicity: that partial reflection of light on transparent surfaces existed: that all this was shown by experiments, but that the corpuscular theory could not explain all this by a reasoned action between matter and luminous corpuscles. Instead Newton wished to attempt a first step towards the theory; so he introduced the definition of 'fits' and of the 'interval of fits'. A very dramatic page follows:

> What kind of action or disposition this is; whether it consists in a circulating or a vibrating motion of the ray,[21] or of the medium, or something else, I do not here enquire. Those that are averse from assenting to any new discoveries, but such as they can explain by an hypothesis may for the present suppose that as stones by falling upon water put the water into an undulating motion, and all bodies by percussion excite vibrations in the air, so the rays of light, by impinging on any refracting or reflecting surface, excite vibrations in the refracting or reflecting medium or substance, and by exciting them agitate the solid parts of the refracting or reflecting body, and by agitating them cause the body to grow warm or hot; that the vibrations thus excited are propagated in the refracting or reflecting medium or substance, much after the manner that vibrations are propagated in the air for causing sound, and move faster than the rays so as to overtake them; and that when any ray is in that part of the vibration which conspires with its motion, it easily breaks through a refracting surface, but when it is in the contrary part of the vibration which impedes its motion it is easily reflected; and, by consequence, that every ray is successively disposed to be easily reflected, or easily transmitted, by every vibration which overtakes it. But whether his hypothesis be true or false I do not here consider. I content myself with the bare discovery that the rays of light are, by some cause or other, alternately disposed to be reflected or refracted for many vicissitudes.[22]

[21] This is a reference to the theory of colours put forward by Descartes and by Grimaldi.

[22] Op. Cit. p. 493

Newton had lost the battle. He himself had provided the weapon which destroyed his own theory. Just as Grimaldi discovered the phenomenon of diffraction while searching for a material light, so Newton was led to study the laws of interference while he was attempting to include his 'corpuscles' in the imposing framework of universal gravitation. Fate could not have been more unkind to both of them. It is clear that Newton must have realized the difficulty of the situation, but he was not prepared to surrender, thus, in the pages discussed, there are many incongruities which are unworthy of him. For example, Proposition XIII came within a few pages of Proposition XII, with which it disagreed; at the same time, it lent itself to an immediate and unavoidable criticism. If the partial reflection of transparent surfaces was due to the presence in the luminous rays of 'fits of easy reflection or easy transmission', why was it that fifty per cent of the incident light was not reflected? Why should the 'disposition' to reflection vary with incidence? It is unlikely that Newton did not ask himself these questions, but he knew that the answers would have been fatal to his theory, and so he preferred not to mention them. In this frame of mind he faced the third part of his work, and came up against two formidable obstacles, namely diffraction and double refraction.

The third Book of *Opticks* is not of the same standard as the previous two. With our present knowledge of the phenomenon of diffraction, we can easily criticize the content of Book III, both from the experimental and theoretical point of view. It is true that we must admire the method and the acumen with which experiments are carried out on prismatic colours and on the rings, even if in the theoretical treatment we can point out details of particular importance which Newton seems to have neglected. When, on the other hand, we make a critical examination of the group of experiments concerning diffraction, it becomes obvious that he was so biased that a free interpretation of the experiments was not possible. (Fig. 70.)

The first title of Book III: 'Observations concerning the inflexions of the rays of light, and the colours made thereby', immediately shows the intentions of the author. The 'inflections'[23] to which he alludes are the phenomena of diffraction. In this he was determined to see an effect similar to that of reflection and he did not even mention the name, diffraction, proposed by Grimaldi for this phenomenon, because he did not wish to mention Grimaldi's view that light could be propagated in four ways: directly, by reflection, by refraction, and by diffraction.

In the frontispiece of his work Newton mentioned '*de reflexionibus*, *refractionibus*, *inflexionibus et coloribus lucis*', as if he wished to indicate that 'inflections' were something different from 'reflections'. Since the author himself did not choose to give an explicit definition of the meaning of the word 'inflection' we must deduce that it is a question of a phenomenon similar to reflection, but not

[23] Concerning this word we wish to call the attention of the reader to a very strange case. Since the first English edition (1704) and the first Latin edition (1706) of *Opticks* there have been many published and in all, at least in all those which we have been able to consult, the word used at this point is *inflections*, the only exception being the Lausanne edition (1740) in which we find the word *reflections*.

When we first consulted *Opticks* we studied the Lausanne edition and we drew the conclusions which were stated in the previous editions of this *The Nature of Light* in other languages. It never occurred to us to verify the Lausanne edition, which is excellent, by comparing it with others. Recently Professor R. Savelli called our attention to what he thought was a typographical error. Actually the typographical error exists in the Lausanne edition. However the deductions that we have made and which follow in the text are not affected because the *inflection* introduced by Newton, is in reality a *reflection* even if a particular type of reflection. As we have pointed out in connection with figure 69, Newton did not consider reflection as the effect of an elastic collision of the particles of light, with the surface of the reflecting body, but rather as a bending of the trajectory of the particles of light, similar to that which rays undergo in the case of a mirage. The upper part of figure 70 shows that Newton made the corpuscles of light deviate just as if they were reflected by the obstacle, according to this concept of reflection, namely as if the corpuscles underwent a repulsion by the obstacle, with more marked effect as the corpuscles approached the material surface. Even if the word has been changed, substantially the idea is the same.

ruled by the law of the equality of the angles of incidence and reflection. Had he admitted that there existed a fourth way of propagation of light, namely by diffraction, he would have had to find another mechanical attribute in his model to explain it. But as this was not acceptable in his model, he attempted to include diffraction in the group of phenomena of reflection and refraction already known, and his experiments followed this line. Unfortunately in spite of his efforts the experiments did not give him the results he desired.

Newton not only repeated Grimaldi's experiments but he extended them and added many more measurements. He carried out experiments in white light as well as in monochromatic light produced by a prism; he observed the diffraction phenomena produced with a slit of adjustable width, and finally he carried out the experiment with a slit having its edges inclined to form an angle. In figure 3 of Table I, Book III, the phenomena are beautifully reproduced, and among them are shown the characteristic hyperbolic fringes, corresponding to the apex of the slit. Among the measurements which interest us the most are those relating to the distance between the diffraction fringes which varied as the distance between the diffracting obstacle and the plane of observation varied. These measurements, carried out with Newton's usual skill, showed that the fringes themselves did not belong to planes which were tangential to the obstacle, but rather belonged to curved surfaces, as Grimaldi had already discovered. This must have been a source of serious difficulty for Newton. (Fig. 70.)

Another important observation was that relating to the dark band which appeared in the middle of a beam of light when it passed through a slit placed at a suitable distance, the band being a function of the width of the slit itself. Newton not only noticed this dark band, but he also noticed that it could be made to appear and disappear by varying the width of the slit while keeping fixed the screen on which it was observed. Alternatively the same effect was obtained by maintaining the same width of slit while moving the screen closer or further away. This must also have caused Newton some embarrassment, judging by the fact that he did not give many details of this experiment.

Contrary to the previous Books, in which he first described objectively and clearly the observed phenomena and then discussed the theoretical interpretation, he now described the experiments in terms of his theory, as if he wished to induce the reader to accept them only in this form. For example, in Observation 6, after placing two knives with their edges facing each other so as to form a slit, he continued:

> . . . and when the distance of their edges was about the 400th part of an inch, the stream parted in the middle, and left a shadow between the two parts. This shadow was so black and dark, that all the light which passed between the knives seemed to be bent and turned aside to the one hand or to the other.

These few words indicate how Newton was attempting to explain the diffraction produced by a slit. The edges of the knives 'inflected' the light as if by a kind of attraction or repulsion, and divided the light into two beams, so that on the screen, at a given distance from the slit, a dark line appeared which was parallel to the edges of the slit. Newton stressed the fact that the rays passing through the middle part of the slit formed the central part of the figure on the screen; but he did not mention the disturbing fact that when the screen was moved further from the slit, then light took the place of the dark line. There was also another inescapable conclusion which would present itself to anyone trying to follow Newton's reasoning. Since the diffraction fringes did not belong to a plane, but rather to a curved surface, then the action of attraction or repulsion of the edges of the diffracting object should continue even when the corpuscles were far away.

Diffraction was a strange phenomenon. If light was allowed to fall on a human hair, the shadow produced was wider than that predicted by geometrical optics; if light was allowed to pass through a slit, the beam instead of becoming narrow was found to be wider. Even if we are prepared to ignore the small fringes which are always present at the edges of these beams and shadows, in the first case we could imagine them as Newton did, caused by a kind of reflection at the edge of the obstacles grazed by light. The same rule could not be applied to the second case. This suggested, as Newton himself thought, the existence of an inflection in the opposite direction to the earlier reflection. The difficulty was to reconcile these two observations. Newton did not even attempt it, but he only sketched the outline of an explanation, which can be seen in a Scholium in the *Principia*. When light reached a body that had the shape of a wedge, such as the edge of a knife, part of it was refracted in the same way as when it passed through a prism, while part of it went round the edge of the knife and produced an inflection. Later Newton often used this model, forgetting the earlier model he had suggested to explain the widening of the dark shadow. It is clear that Newton was familiar with Grimaldi's work, because he mentioned him as the discoverer of the phenomenon, though he did not use the word 'diffraction'. Yet he seemed to forget that Grimaldi had used many arguments and experiments to prove that in diffraction both reflection and refraction were to be excluded. Newton limited himself to the description of diffraction by experimenting with knives and sharp edges; he mentioned the circular and rectangular edges of gold, silver and brass coins, specifying the '*termini rectanguli circulares*' as if when the coins were placed edgeways the phenomenon did not appear. He also quoted the edges of knives, of broken pieces of stone or glass, and he only used human hair, threads, and needles when he wished to obtain the 'inflection' which widened the shadow. Why did he not study diffraction using a slit formed by two needles rather than two knife edges? This type of experiment should have been carried out, particularly since other scientists had already noticed that Newton's claims were not all he wished them to be. Newton, however, knew that he was not on the right track. In his writings the man took over from the scientist, and did his best to obtain some sort of explanation of these strange and disconcerting phenomena. After this modest attempt he appeared to withdraw in good order, and he ended his description of Observation II with these words:

> When I made the foregoing Observations, I designed to repeat most of them with more care and exactness, and to make some new ones for determining the manner how the rays of light are bent in their passage by

> bodies, for making the fringes of colours with the dark lines between them. But I was then interrupted, and cannot now think of taking these things into further consideration. And since I have not finished this part of my design, I shall conclude with proposing only some queries, in order to a further search to be made by others.[24]

Newton, therefore, entrusted others with the task of explaining the phenomena which he realized he could not explain. There is no doubt that this was a clever way out of the difficulty and the consequences have been those Newton probably expected.

Newton summarized his experiments and theories on the nature of light in 31 'Queries', which take sixty pages. Particularly interesting is 'Query 25' where he deals with double refraction. Before we discuss this other serious obstacle met by Newton, we should note some peculiarities which are contained in the 'Queries' but which are not discussed in the earlier pages of the work. In the case of diffraction, in spite of all the reflections and inflections, it was still difficult to understand why the light reflected by thin threads or inflected by the edges of knives, persisted in appearing as parallel fringes at the edges, and was not propagated in a straight line. Newton advanced an idea, but so hesitantly as to put it forward only in the form of a 'Query'. In fact, 'Query 3' stated:

> Are not the rays of light, in passing by the edges and sides of bodies, bent several times backwards and forwards, with a motion like that of an eel? And do not the three fringes of coloured light above mentioned arise from three such bendings.[25]

This was a bold suggestion, but there is no point in insisting on diffraction since Newton had already dismissed it.

In 'Query 12' and '13' there is an interesting note concerning vision. Newton asked if we should not consider the possibility that the corpuscles, in falling on the back of the eye, excited some vibrations in the retina, and that these, in being propagated along the

[24] Op. Cit. Book III, Part I, p. 516

[25] Op. Cit. p. 516

optic nerve until they reached the brain, produced the sense of sight (*sensum videndi*). He added that rays of different types should excite vibrations of different magnitude (*diversa magnitudo*) and according to this different magnitude produce sensations of different colours, just as it happened in the case of sound. Immediately after this he specified that the vibrations relative to the 'most refrangible rays' namely violet rays, should be 'the shortest' and those relative to the 'least refrangible rays', namely the red rays, should be 'the largest'. In this question we can detect the influence of Newton's predecessors, especially that of Grimaldi. In the pages which followed, Newton, however, specified his ideas in a different way. He is led to consider the existence of a medium more rarefied and more elastic than air, which was present in the vacuum, in all bodies and in the whole universe and which, by means of its vibrations, transmitted heat, and he called this medium 'aether'. The transmission of the luminous stimulus from the eye to the brain through the optic nerve, took place by means of the vibrations of this ether.[26]

We now come to 'Query 25' where the subject of double refraction is only outlined. Newton reported several experiments, some that had been carried out by Erasmus Bartholinus, and by Huygens, and some original ones which he performed himself; but when he came to the point of giving an explanation and of coordinating it with the ideas he had until now staunchly supported, he limited himself to asking questions, leaving the answers to the future investigators who would deal with this subject. The phenomenon of double refraction is well known. When a beam of light, even if monochromatic, falls upon the plane surface of a crystal, especially of Iceland spar, but also of quartz or of other substances, it divides into two beams. One of these follows the common laws of geometrical optics and is called the 'ordinary' beam, while the other, called the 'extraordinary' beam, follows laws which are completely different. This extraordinary beam, except in particular circumstances, does not even remain in the plane defined by the normal to the surface of incidence and by the incident beam. If the

[26] Op. Cit. Query 23 and 24 p. 522

FIG. 73. Christian Huygens (1629–1695)

TRAITE
DE LA LVMIERE.
Où sont expliquées
Les causes de ce qui luy arrive
Dans la REFLEXION, & dans la
REFRACTION.
Et particulierement
Dans l'etrange REFRACTION
DV CRISTAL D'ISLANDE.
Par C. H. D. Z.
Avec un Discours de la Cause
DE LA PESANTEVR.

A LEIDE,
Chez PIERRE VANDER AA, Marchand Libraire.
MDCXC.

FIG. 74. Frontispiece of *Traité de la Lumière* by Huygens

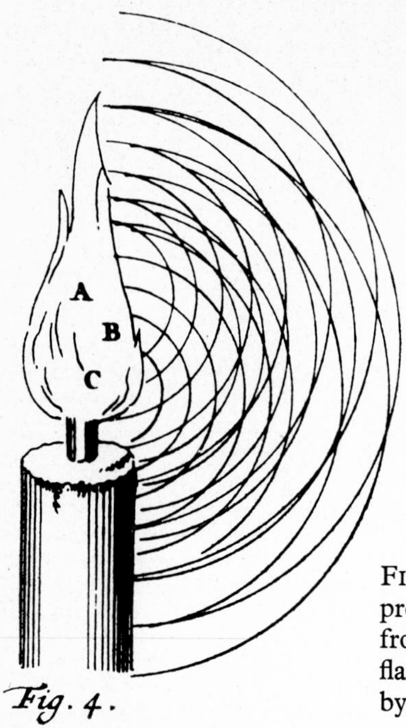

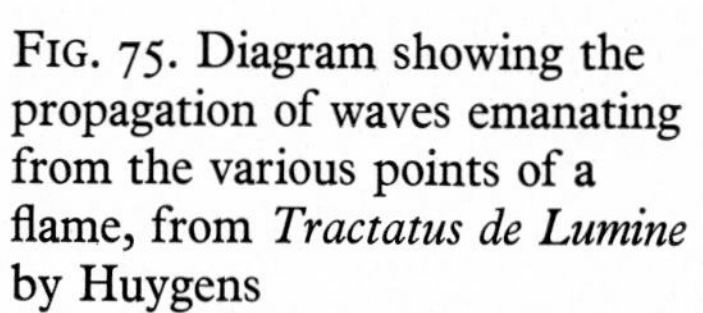

FIG. 75. Diagram showing the propagation of waves emanating from the various points of a flame, from *Tractatus de Lumine* by Huygens

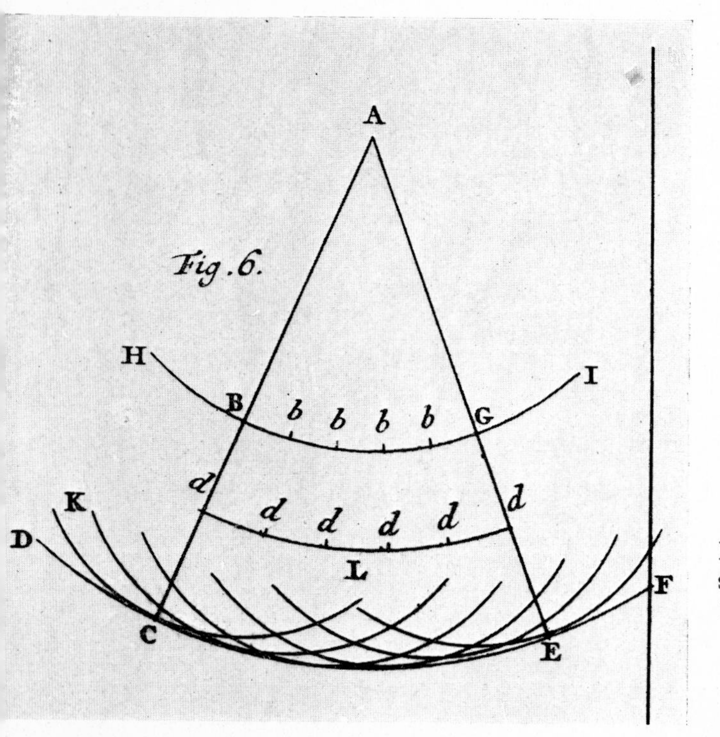

FIG. 76. Envelope of secondary wavelets, from *Tractatus de Lumine*

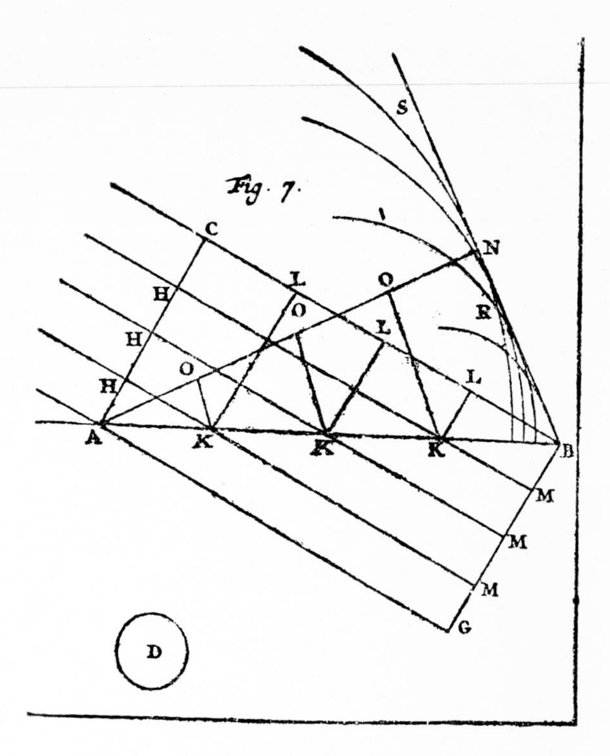

FIG. 77. Diagram showing the reflection of light from plane aces, from *Tractatus de Lumine*

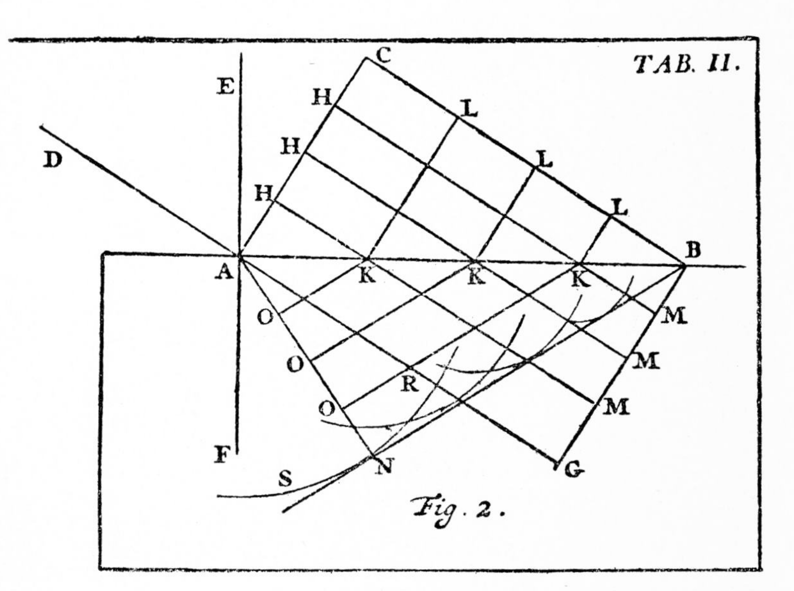

FIG. 78. Diagram showing the refraction of light through a plane surface when the velocity in the second medium is less than in the first, from *Tractatus de Lumine*

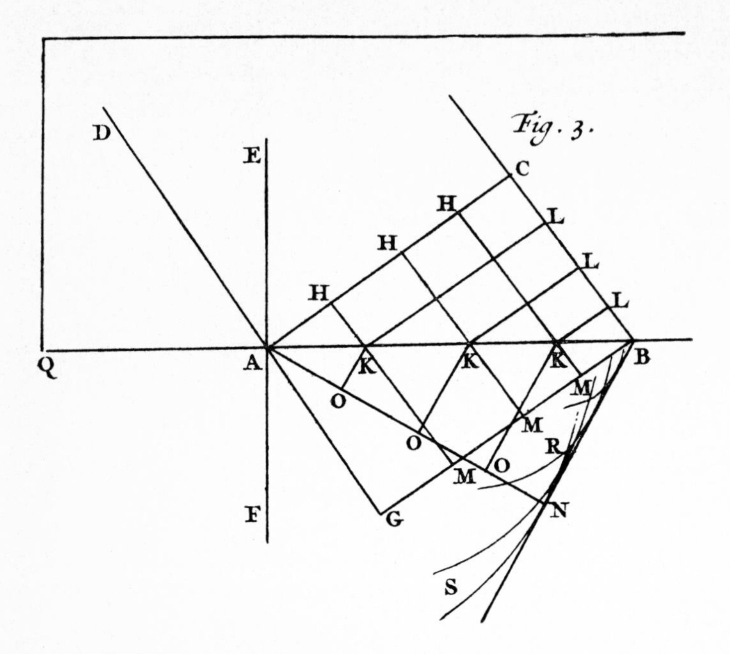

FIG. 79. Diagram showing the refraction of light through a plane surface when the velocity in the second medium is greater than in the first, from *Tractatus de Lumine*

FIG. 80. Reproduction of an original drawing by Huygens intended to represent the envelope of refracted secondary wavelets, from *Oeuvres complètes de Christian Huyghens*

two beams which are produced by a crystal of Iceland spar are allowed to fall upon another crystal which produces a similar phenomenon, then each beam, generally, is split again into two new beams with complementary intensity; and for certain orientations of the new crystal, the two beams, although they are refracted, are not doubled any longer, but remain as two. This is true even if the second crystal is rotated by 90° around the axis determined by their direction, provided that we start from a position when this condition already existed. There is no doubt that this phenomenon must have been very disturbing for anyone trying to see in the phenomena of reflection, of refraction and of inflection, the effect of one single action of matter upon the corpuscles.[27] Newton believed that he had made a discovery when he proved that reflection was not due to the collision of luminous corpuscles with the material parts of solid bodies, but rather was the result of a general effect due to all these parts, inasmuch that a corpuscle was attracted more by the matter of one medium than by that of the medium which was on the other side of the reflecting surface of separation. Already, however, the partial reflection had demolished the whole construction, leaving the luminous particle in these circumstances, free to decide, as it were, whether to be reflected or refracted. As if all this was not enough a new complication now arose. When the corpuscular rays met the transparent surface of one of these substances like Iceland spar, which today we call birifrangent substances, they were partly reflected in one direction (and no one remarked on this), and partly refracted, some following one law and others following another. Each of the two groups maintained properties which were different from the other, indeed these properties were almost complementary to each other, and they were characterized by two planes of symmetry at right angles to each other and containing the direction of propagation. All this was very confusing indeed, and Newton had to admit the existence of a new characteristic which was inherent in the corpuscles themselves, and as a result of which the crystal acted as a detector or rather as a discriminator. He concluded 'Query 25' by stating:

[27] Op. Cit. Query 4 p. 516

'. . . The unusual refraction is, therefore, performed by an original property of the rays. . . .'[28]

In the following 'Query' Newton asked himself whether this could possibly be due to the special form of the sides of the corpuscles. This hypothesis may appear attractive, but it is not possible to ignore the fact that the effect of a dissymmetry of the corpuscles can only originate by contact, namely by a collision with material particles; and it is impossible to see how this could agree with the previous model. On this point Newton is silent. He only advanced a vague idea, giving as an example small magnets which have 'privileged' poles or points. He also suggested the possibility of a dissymmetry within the particles forming the birifrangent crystal, but he did not give any details of the mechanism which could explain a deviation that would produce the strange effects which are observed in the phenomenon of double refraction. Indeed it is not easy to explain the fact that when a beam of corpuscles, even if individually dissymmetrical, falls perpendicularly on a plane surface, part is reflected normally, part continues inside the crystal without bending, and part assumes a different direction inside the crystal itself. Newton left the solution of this complex phenomenon to posterity.

We have stressed the weak points of Newtonian thought because we wished to reach a conclusion that certainly was not common in the eighteenth century. Newton never tried to assert anything in a definite manner about the nature of light. Many thought that they detected in this a philosophical principle of someone who did not wish to put forward theories, but rather wished to take the experimental reality as a self-sufficient truth, in other words a positivism openly professed. For our part, we see in this a great effort to reach a theoretical construction of far-reaching effect which was followed by a failure personally felt, but not openly admitted. During the first period of his brilliant experimental research, Newton had great hopes. But he had more or less openly to gloss over a second group of phenomena which refused to be included in the general

[28] Op. Cit. p. 524

picture as he saw it; finally he had to capitulate when faced with other phenomena which were definitely irreconcilable. He realized that it was impossible, because of the nature of his construction, to include all the optical phenomena that were then known, even by making mental sacrifices, theoretical acrobatics, and philosophical renunciations. Nevertheless, because he had collected a great deal of data in his attempt to reach his goal, he continued his efforts in the hope that his successors could reach that goal, or at least, collect further data in their attempt to do so. He did this with great skill. He never expressed himself decisively in favour of any particular theory of the constitution of light. All this is very evident when we consider the technique he used, the experiments he made, and the description he gave of these experiments. Whoever reads Newton's *Opticks* without prejudice, without blind admiration, and with a critical and clear approach, must conclude that no one could have worked better than Newton, not to build, but rather to demolish, the corpuscular theory.

It is amazing, we could almost say inexplicable, how a construction which was so disconnected and insecure, full of incongruities, and of omissions, could have convinced the majority of the physicists of the eighteenth century, and could have been accepted even beyond the technical field, as a great scientific conquest. Newton's fame was so great that many ideas were attributed to him that he would himself never have been willing to support, since he knew only too well that they were unsustainable, yet there was no one who dared even to doubt them.

The characteristics of light according to Newton can be summarized as follows:

Light consisted of extremely small and fast corpuscles called rays of light, which followed a rectilinear propagation in any homogeneous and transparent medium, without being affected by matter or by nearby rays.

Rays of equal dimensions were homogeneous, and when they struck the back of the eye, they produced vibrations which were transmitted to the brain, where they stimulated the sensation of colour, through the 'ether' of the nerves.

Rays which had different 'sizes' produced different colours.

The larger ones produced the colour red and excited at the back of the eye the larger vibrations. The smaller rays produced the colour violet and excited shorter vibrations at the back of the eye. Intermediate rays had intermediate size and produced the intermediate colours of the spectrum.

Rays had special forms with a polarity which was comparable to that of small magnets, and had two planes of symmetry at right angles to each other, passing through their trajectory.

The rays had different velocities in the various diaphanous media. They had a minimum velocity in vacuum, and the velocity increased with increasing density of the medium in which the rays moved. The increase in velocity was greater for the violet rays than for the red.

During rectilinear propagation the rays were also endowed with an oscillating property which could be produced by a wave that came into being at the same time as the corresponding ray, and which travelled in the ether preceding the ray itself. The ether was an elastic and very rarefied substance which pervaded the whole universe, including material bodies.

When the rays collided with material bodies they could be absorbed, but at the same time these bodies were agitated and became heated.

When rays met a surface of separation between two media, they could either penetrate or not penetrate it according to the attraction that the matter exerted upon the rays, the angle of incidence, and the actual conditions of the oscillating property just mentioned, which today we would call the 'phase'. Generally this property was such that the majority of the rays penetrated the body and only a very small percentage, which varied with the angle of incidence, was reflected. The hypothetical waves which preceded the rays followed the same fate as that of the ray with which they were associated.

When the rays were refracted by penetrating a new diaphanous body, they were bent according to their dimensions. As a result the directions of propagation of the corpuscles of different size were separated in relation to the diversity of the dimensions and the conditions of incidence.

When these rays met opaque bodies they were partly reflected and partly absorbed according to their dimensions. In this way they produced the colours of the bodies.

When these rays met very thin films, their reflective and refractive properties were accentuated. Zones of full reflection and of full transmission were produced according to the size of the rays and the thickness of the film, there being a definite correspondence between 'refrangibility' and the width of the 'intervals of maximum reflexibility'.

When these rays passed near hairs, threads, needles, and other similar objects, they were reflected by these and produced shadows which were wider than the geometrical shadows. This inflection by small obstacles was accompanied by a triple pulsation which had an undulatory motion like an eel; the rays lost the property of travelling absolutely in a straight line and divided themselves into series of three fringes or bands.

When these rays grazed the edge of a knife or a similar object, they were bent towards the knife itself and produced phenomena similar to those described above.

When these rays met the surface of a crystal of Iceland spar, or quartz, or some other substance, a few of them were reflected in one particular direction according to the well-known law of reflection, while the other rays were refracted in two directions, which generally were different. One group of rays followed the ordinary laws of refraction, and the other group followed a completely different law. The rays which travelled through these crystals maintained a certain symmetry and special properties with reference to two planes at right angles to each other, passing through the trajectory of propagation.

In the above we have a very simple summary of the model of light attributed to Newton. The value of it is the description of some remarkable experiments in optics: a demonstration that the corpuscular model is insufficient to represent the known optical phenomena, and in addition the necessity of making use of an undulatory theory. No one had shown this better than Newton, and yet while he very ably and with great sincerity never gave decisive support to the corpuscular theory, he definitely opposed the undulatory theory, at least judging by his own writings. He continually found great difficulties in using the corpuscular theory, and was himself increasingly impelled towards the undulatory theory, when faced with the transmission of the stimuli from the retina to the brain, and the structure of 'fits of easy reflection or of easy transmission'. However, he staunchly refused to accept the propagation of light in diaphanous bodies by means of waves, because he could not reconcile it with rectilinear propagation. This seems even more strange when we remember that he had been compelled by the phenomena of diffraction to reject, even if reluctantly, the absolute rectilinear propagation of corpuscles. His aversion to the propagation of light by waves appears in his 'Query 28' when he asked: 'Are not all hypotheses erroneous in which light is supposed to consist in pression or motion, propagated through a fluid medium . . .?' Perhaps all this is even clearer in Section VIII of Book II of *Principia*, entitled 'The motion propagated through fluids'. Here Newton studied the propagation of waves in relation to the period, the wavelength, the velocity, the density, and elasticity of the medium. Even today these studies are considered correct, and are generally accepted. His intention is shown simply by a drawing (fig. 72), which shows that shadows could not be produced by undulatory propagation.

Section VIII of Book II of *Principia* ended with a 'Scholium', in which the conclusions expressed in the same Section and obtained in the field of mechanics are extended to light and sound. The text is as follows:

> The last Propositions respect the motions of light and sounds; for since light is propagated in right lines, it is certain that it cannot consist in action alone (by Props. 41 and 42).

There is no further mention of light in relation to undulatory propagation. The two Propositions quoted above stated: Proposition 41, Theorem 32:

> A pressure is not propagated through a fluid in rectilinear directions except where the particles of the fluid lie in a right line.

Proposition 42, Theorem 33:

> All motion propagated through a fluid diverges from a rectilinear progress into the unmoved spaces.

These two Propositions are extremely interesting, because the *Principia* was published in 1686, eight years after Huygens had prepared and made public, but not published, the text of his *Traité de la lumière*. This little book was printed in 1691, but had been written in 1678. Huygens, in the preface, stated that he had read the book to many members of the Paris Académie des Sciences, among whom he mentioned the names of Cassini, Roemer, and De la Hire. In any case, the opinion expressed very clearly by Newton about the undulatory propagation was taken up again in 'Query 28' of *Opticks*, which was published in 1704, that is to say thirteen years later than Huygens' work. Indeed in this particular work Newton quoted the explanation that only Huygens had attempted to give of double refraction by means of the undulatory theory. He quoted it to confirm his own argument; and he gave the original sentence of the text with which Huygens admitted that he could not fully account for such a strange phenomenon. This is the only quotation of Huygens to be found in the whole of *Opticks*.

It is curious to compare the fate of Newton with that of Grimaldi. In the work of both the desire to prove the material and corpuscular nature of light is evident, and yet both these men were led by their own studies to admit undulatory characteristics in their models of light. Both left their names connected with phenomena typically undulatory: the diffraction of Grimaldi and the rings of Newton. It is not known exactly where, and by whom, the corpuscular theory was created. Yet it was considered to be not only satisfactory, but wonderful during the whole of the eighteenth century. Perhaps the simplicity with which Newton's theory explained the best-known elementary phenomena of light, such as reflection, refraction, and the production of colours, conquered the majority of minds. This left unexplained the complex phenomena of diffraction and double refraction, which waited for future scholars to include them in the general framework of a simple and acceptable theory.

This trend of thought existed most in the minds of ordinary and superficial thinkers, but there were wiser minds that were not prepared to accept uncritically all that the less discriminating readers attributed to Newton. Foremost among these was Huygens.

Christian Huygens, born at The Hague in 1629, was thirteen years old when Galileo died and twenty-one when Descartes died. From these two great men, particularly from Descartes, he assimilated the doctrine and the new philosophical principles of the science of nature. His studies were mainly directed towards the experimental approach, and in many respects he reached technical perfection. In the field of theoretical speculations, he was very demanding and it is enough to read what he wrote clearly in the first pages of his *Traité de la lumière*:

> ... at least in the true Philosophy in which the causes of all natural effects are conceived in terms of mechanical motions, this, to my mind is necessary if we do not wish to renounce all hope of ever understanding anything in Physics.[29]

His work is very extensive and devoted mostly to mechanical and optical questions, but there is a comparatively small book of only ninety pages which deals with the question of the nature of light. Even so, the work on this small book dragged on for many years, and in the end was deliberately left unfinished. In the Preface to the first edition in 1691, published in French, Huygens himself related that he had written this small treatise twelve years earlier, while in France, and that he had communicated the content to the Académie des Sciences in 1678, as we have already mentioned above. He asserted that he had not made any substantial amendments to the original, but that he had added only his ideas on the constitution of Iceland spar and on the discovery of the

[29] C. Huygens, *Tractatus De Lumine* in 'Opera Reliqua' Waesberg, Amsterdam 1728, Vol. I p. 2

Also *Oeuvres complètes de Chr. Huygens*, Martinus Nijhoff La Haye, 1937, p. 541

phenomenon of double refraction by quartz crystals. Huygens himself volunteered an explanation for the delay in publishing his work. He explained that he had written his book in poor French, and wished to translate it into Latin, to add it to his treatise on instrumental optics; but in the end had decided to publish the work as it was, so that it should not be lost to posterity. He wrote this at The Hague in 1690, at the age of sixty-one, after delaying the publication for twelve years. He concluded the short preface by expressing his sincere hope that others would follow the study of the subject according to the lines laid down by him, because he was convinced that this field of studies was far from being exhausted, not only because he thought that some of the problems were still unsolved, but also because there were many more which he had not even studied. In fact he did not deal at all with luminous sources and with colours; indeed in the whole of his work the word 'colour' is only mentioned in the preface. He avoided altogether words which might allude to the concept of colour; even when, in discussing technical questions such as the shape to be given to lenses to obtain better defined images, he wished to allude to chromatic aberration, which Newton had fully discussed when pointing out the advantages of reflecting telescopes. Thus Huygens used circumlocution:

> ... because there is a certain property in refraction itself, which presents the perfect crossing of rays, as Newton has so very well proved by experiment. . . .[30]

Huygens, therefore, unlike Grimaldi and Newton, did not face fully the question of the nature of light. Nevertheless, in a few pages of his work are found some fundamental concepts which

[30] Op. Cit. p. 77. It is interesting to see what he wrote to Leibnitz on August 24th, 1690, when sending him one of the very first copies of the *Traité de la Lumière*

> ... Je n'ay rien dit des couleurs dans mon *Traité de la Lumiere* trouvant cette matière tres difficile; sur tout à cause de tant de manieres differentes dont les couleurs sont productées. *Oeuvres complètes de Chr. Huygens*, 1901, Vol. IX Correspondence No. 2611, p. 470

were not followed up and developed at the time, perhaps because the originator himself did not concentrate on them as much as they deserved. These concepts were to become formidable weapons in the hands of the scientists of the following century.

In our analysis of the work of Huygens we will refer to the Latin edition *Tractatus de Lumine*, which was published in 1728, and which was the translation of the French treatise of 1691 that we have already mentioned.

The book consists of six chapters. The first deals with the propagation of light, the second with reflection, the third with refraction, the fourth with atmospheric refraction, the fifth with the 'wonderful refraction' produced by Iceland spar, and the sixth deals with questions of practical optics. Like his predecessors, Huygens began his work by showing how insufficient the existing theories were. He found it strange that theories so badly expressed should have been accepted as true and proved, for example, the fundamental idea concerning the rectilinear propagation of light or that concerning the crossing of rays which did not interfere with each other. It is interesting here to note how Huygens began his attack on his adversaries by turning against them the very weapon which they were using themselves against the wave theory, namely the acceptance of the presumed rectilinear trajectory of light. Huygens could not accept that corpuscular light could penetrate matter without, at the same time, undergoing some sort of disarray and diffusion. He did not consider satisfactory the models advanced by Descartes, Grimaldi, and Newton to explain and to justify how this could happen. 'There can be no doubt that light consists of movement of certain matter'. This assertion could not be more definite. It is in great contrast with the reservation and caution of the supporters of the theory of material light. What could have been the reasons for such certainty? According to Huygens, light on earth was mainly produced by fire and by flames, each of which was moving rapidly; and if light was concentrated by means of a concave mirror, it burnt like fire, that is to say it disintegrated the parts of an object and this: '. . . surely indicates motion, at least in "true Philosophy",' that is Philosophy as he understood it. Moreover, the fact that to obtain vision it was necessary to stimulate the ends of the nerve at the back of the eye, agreed, according to Huygens, with the idea that light was the movement of the matter existing between the object seen and the eye itself. According to the principles of mechanics, which make wide use of the composition and resolution of motion, two or more motions which are propagated in different directions cross and overlap without disturbing each other, but it is impossible to understand how this can take place in the case of swarms of material particles. Finally the phenomenon of light must be similar to that of sound. How often was this parallel to be made in twenty centuries!

Huygens felt that all these arguments were sufficient for him to conclude that:

> . . . there is no doubt that light also comes to us from a luminous body by some motion impressed on the matter in-between, since, as we have already seen, this cannot be by the transport of a body which passes from the luminous object to us.[31]

Indeed if we consider all that had been written on the subject by the predecessors of Huygens, we must admit that notwithstanding his 'true Philosophy', he had been very quick, perhaps too quick, in reaching his conclusions. This confirms what we have already said, namely that Huygens had not given enough time or attention to this subject. On the other hand, the question whether light was motion or matter had been much debated; it had become so confused on account of all the arguments for and against, that the acceptance of either solution was mainly subjective, according to the value attributed to certain arguments rather than others. Huygens preferred motion to matter. Having taken this decision he embarked on the task of formulating a theory which encompassed the fundamental optical phenomena. According to Huygens, the existence of the speed of light denied by Descartes, but which Roemer had determined in 1675, was an argument in favour of his views. From this he went on to discuss the parallel between light and sound. This parallel had always been weakened

[31] Op. Cit. p. 3

by the fact that sound had no shadows, while light (for those who were not prepared to accept diffraction) persisted in travelling in straight lines. The parallel became even more difficult in the second half of the seventeenth century after the invention of vacuum pumps. With their aid it could be shown that sound did not travel in a vacuum, while light did. Huygens was then faced with the problem of the medium in which light was propagated, a problem which had presented difficulties to scientists at all times. Huygens followed this reasoning: sound is propagated in air by virtue of an elastic force or, in other words, by virtue of the reaction to compression, and air consists of material particles that are very small and very mobile. Many well known experiments show that the harder the substance, the higher is the velocity of propagation of elastic waves. Hence to explain the propagation of waves as fast as the luminous waves, it is necessary to admit the existence of an ethereal substance much more tenuous than all matter, capable of penetrating all bodies, of filling all space, and, at the same time, being endowed with a very high degree of hardness and elasticity. He added:

> It is not necessary to examine the cause of this elasticity and of this hardness, the consideration of which would lead us too far from our subject.[32]

Huygens could have found a more plausible excuse. Although his time might have been precious, he could have spared a little to explain this point which was a real mystery to all, especially in those days. He gave the outline of a theory in which he considered ethereal particles. Although they were extremely tenuous, he thought these consisted of aggregates of even smaller particles; elasticity was due to the extremely fast motion of the tenuous matter, which was even capable of penetrating between those particles, and of altering the structure in such a way as to leave an easy passage to fluid matter. Huygens added that in this way, he had extended to the ether theory, Descartes' model of elasticity, though with some improvements. From the manner he presented the whole

[32] Op. Cit. p. 11

issue it is obvious that he did not wish to commit himself further on the subject. In fact he added that we must not be surprised if he postulated a tenuous and elastic substance because:

> though we ignore the real cause of elasticity, we still see that there are many bodies which have this property, so there is nothing strange in attributing elasticity also to very minute, invisible corpuscles like the particles of the aether.

In conclusion, all that is asked of the ether is to be a substance uniformly elastic.

Having thus overcome the first obstacle, Huygens turned his attention to the question of the propagation of waves in the ether. Naturally he had to bear in mind the objection put forward for so long by so many, namely that the propagation in a fluid took place by secondary spherical actions moving in all directions from each point, just like the fire of Grimaldi's example. It is here that Huygens had the brilliant idea which has immortalized his name. He put forward the theory of the envelope of secondary wavelets. The reasoning by which he was led to formulate this concept is of great interest. He gave the example of a flame emitting many individual waves from all its points. Once these waves entered the ether they were propagated independently of each other, because the individual motions were communicated to the various particles of the ether independently both of each other and of other movements which the particles themselves might have, just as sound waves were propagated in air without being affected by the motion of the material particles of the air. Having accepted this, Huygens remarked how extraordinary and almost incredible was the fact that individual waves emitted by such minute sources, could be detected at great distances such as those which separate the stars or the sun from us. How could this be explained? (Fig. 75.)

> We will cease to be astonished when considering that at a great distance from the luminous body an infinity of waves, even if emitted by various points of this body, join together forming one single wave which consequently is strong enough to be detected.

This was the concept of the envelope of waves emitted by the various points of an extended source. (Fig. 76.)

From this Huygens extended the concept to the case of the propagation of a wave emitted by a single point source. It was true that every point where the motion of a particle arrived became a centre of new spherical waves, as Grimaldi had clearly shown, and as Newton had illustrated in the drawings in *Principia*; but each one of these secondary wavelets was extremely weak, and the effect would be visible and detectable only where an envelope of these secondary wavelets was formed, thus becoming a new wave front, which had its centre in the source itself. The wave front here took the place of the flame of the previous example. In this mechanism Huygens saw the key to the explanation of the rectilinear propagation of light in homogeneous media, as well as of reflection, refraction and double refraction. There was even more to it than this, but Huygens did not appear to have suspected the full implications of his theory.

Huygens did not study diffraction even though this would have given the best support to his theory. This is so strange that we decided to investigate the reasons. We have reached the conclusion that it was probably due to a lack of the mathematical knowledge required to deal with the kinematics and the dynamics of periodic motions. In fact, although the concepts of phase and of interference were already beginning to appear in the 'fits' of the corpuscles of Newton, as well as in the various ideas of Hooke and Huygens, knowledge of the kinematics of periodic motions was still very rudimentary; many of the problems appeared so complex to scientists, that they did not even attempt to solve them. For this reason, among others, the concept put forward by Huygens did not have an immediate following. It was too difficult at that time to understand the simplicity of its logic. Even a mathematical mind as powerful as Newton's could not appreciate it, perhaps because it was obscured by preconceived ideas, but more probably because it was untrained in this type of reasoning. The rectilinear propagation of light was still the point most discussed and, at the same time, a point on which opinions were still irreconcilable. Huygens had advanced ideas in favour of a wave theory, and had succeeded in solving the problem for an open arc of wave. His opponents, however, pointed out that had he limited the wave front by means of a diaphragm, the secondary wavelets would have overflowed behind the opaque part of the same diaphragm. To this Huygens rejoined that since it was a question of waves which did not produce envelopes, their effect was negligible and they would not be detectable. Still, the opponents did not agree with this view and Newton's drawings show clearly what he thought of it. When it is a question of deciding whether a quantity is negligible or not, there is only one way and that is either to calculate it or to measure it. At the time no one attempted either, because no one knew how to carry out the integration of so many periodic motions. Faced with this mathematical complication, the discussion was bound to remain in the domain of opinions. Yet the experiment had already been carried out, since diffraction had shown to the followers of the corpuscular theory that light was able to bend a little around opaque obstacles. Although Huygens had the idea that secondary wavelets overflowed behind obstacles, he did not pay any attention to diffraction. Instead he insisted on considering the waves which did not produce envelopes beyond obstacles, as undetectable, without defining whether the zone of the envelope ended with a continuous fading or with a discontinuity. He simply stated: 'And therefore the rays of light can be considered as if they were straight lines'.[33] More than a hundred years had to pass, and the efforts of a physician and of a civil engineer were required before it became possible to find the link between secondary wavelets and diffraction.

Huygens explained clearly the phenomenon of reflection and its laws by considering the envelope of the secondary wavelets which are emitted by the particles of the reflecting surface, as each particle is struck by the incident wave. The drawings in his treatise show very clearly his reasoning which is well known. The same reasoning could also be used to explain refraction only if it was accepted that the velocity of light was different in the two media separated by the refracting surface. The argument really had a

[33] Op. Cit. p. 16

double edge. On the one hand it had the advantage of requiring a smaller velocity in the denser media, which agreed with ordinary common sense, even if it was not agreeable to Descartes or Newton. On the other hand it was necessary to explain why such waves should have different velocities in different media. Were these waves not carried by the ether and did not the ether permeate all bodies? Had it not been stated that an elastic and hard ether was necessary to explain the great velocity of light? How then could matter alter this velocity which depended only on the elasticity and hardness of the medium? (Fig. 77.)

Huygens suggested three possible solutions, but in fact he only succeeded in setting a problem which was to try the greatest minds among his successors. He stated that he did not consider bodies to be like sponges or like risen bread, but like 'a collection of particles in contact and at the same time in such a condition as not to fill space completely'. He then proved that the gaps had to be filled with ether, and he proposed three solutions to explain why bodies were transparent to light.

1. When light passed from a vacuum into a transparent medium, the undulatory motion was communicated from the external ether to the matter of the body. Assuming that this was less elastic than the ether, then the smaller velocity of light in matter than in a vacuum was justified.

2. The undulatory motion continued to be transmitted by the ether which filled the gaps between the material particles of the body, and being compelled to pass through so many narrow channels, the velocity of motion was reduced.

3. The undulatory motion was transmitted both by the matter and by the ether; this was true at least in the special case of bodies which produced double refraction.

We cannot deny that every effort was made to try to solve the problem, but it is also true that the whole argument did not satisfy careful critics. From this model arose a strange difficulty: how was it possible for a body to be opaque? Should there not always have taken place a transmission of elastic waves in matter, or indeed in the ether contained in the gaps between molecules? Perhaps it could be suggested that there was a lack of elasticity in the particles

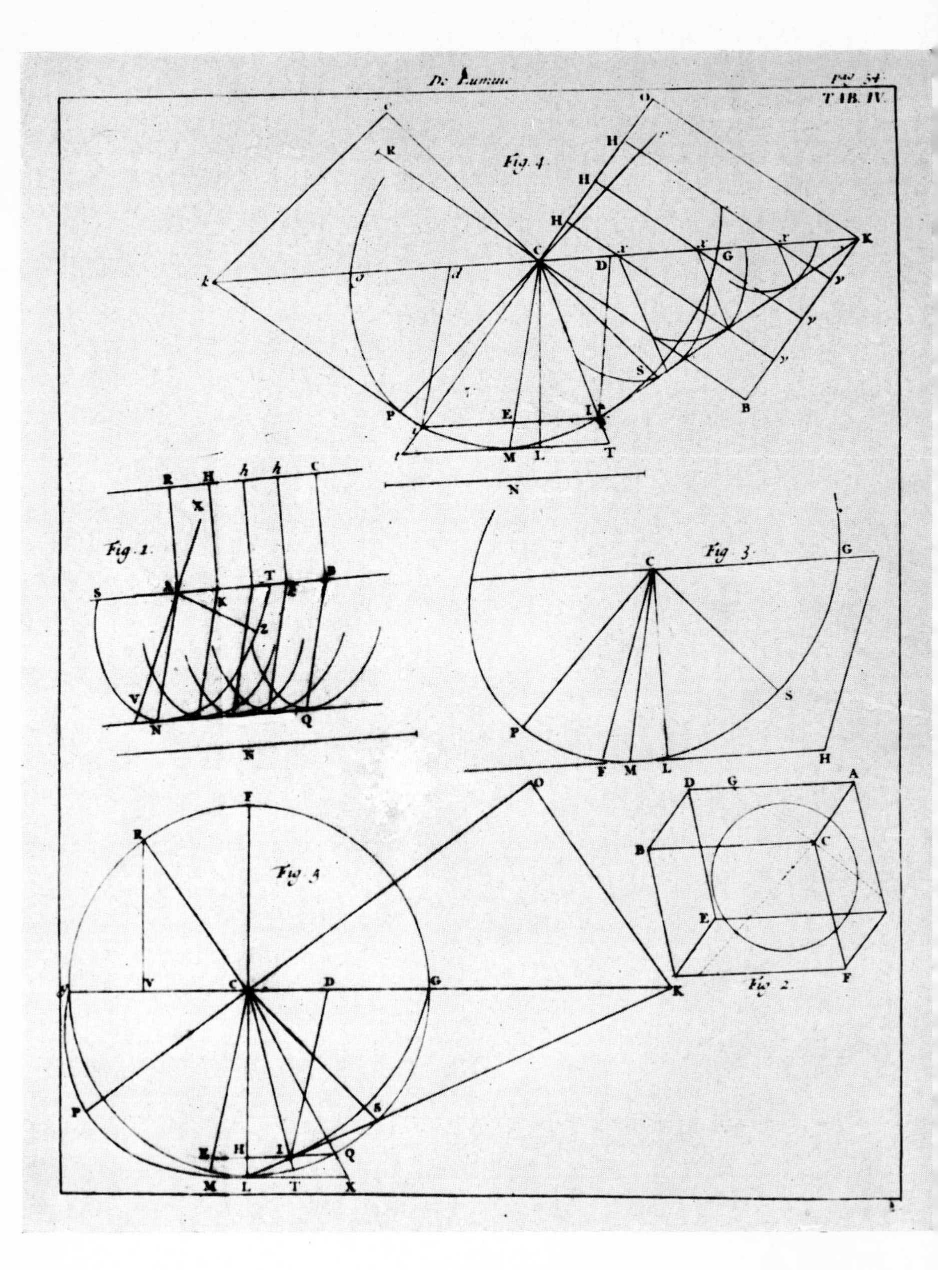

FIG. 81. Diagrams showing the mechanism of extraordinary refraction, from *Tractatus de Lumine*

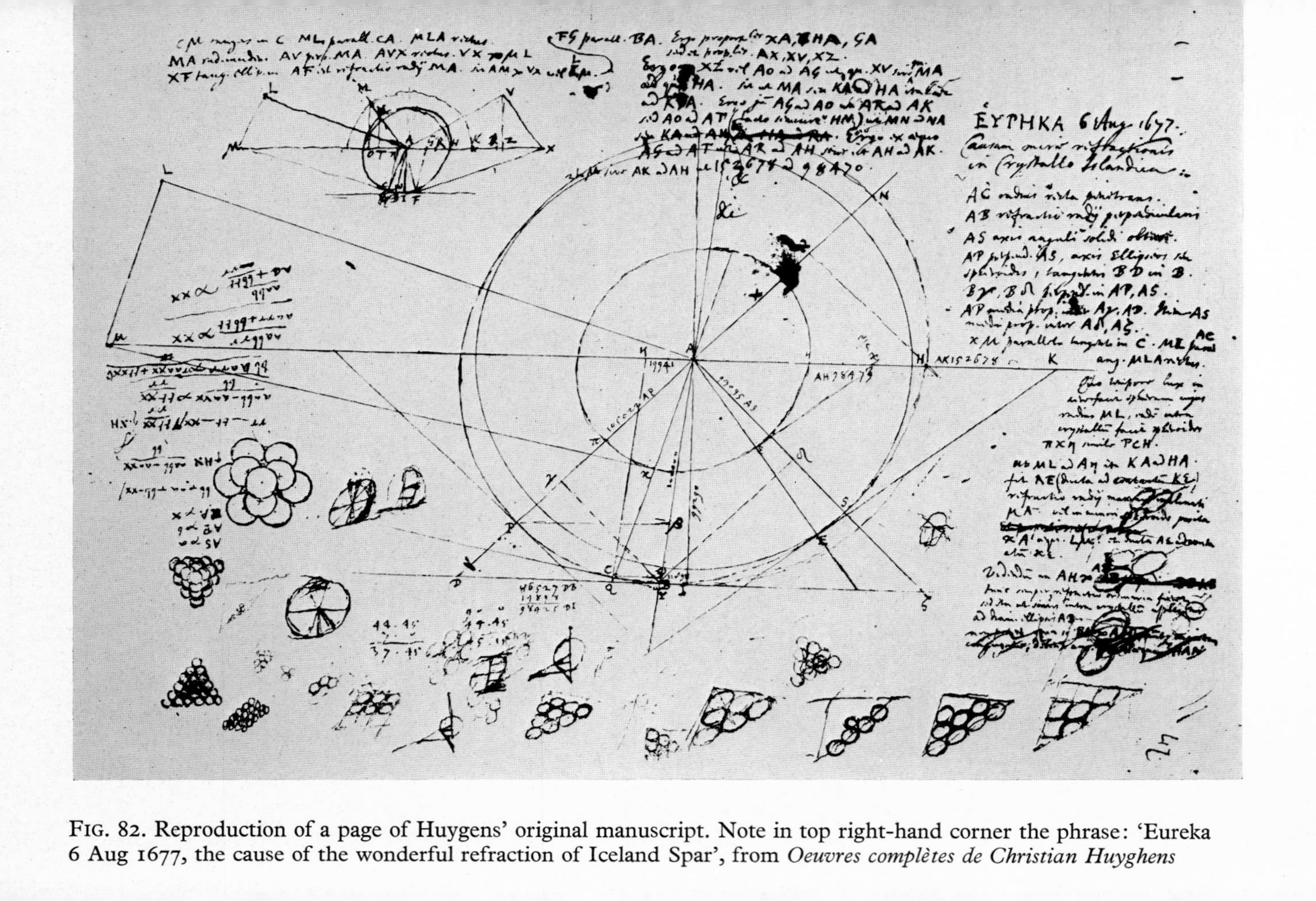

FIG. 82. Reproduction of a page of Huygens' original manuscript. Note in top right-hand corner the phrase: 'Eureka 6 Aug 1677, the cause of the wonderful refraction of Iceland Spar', from *Oeuvres complètes de Christian Huyghens*

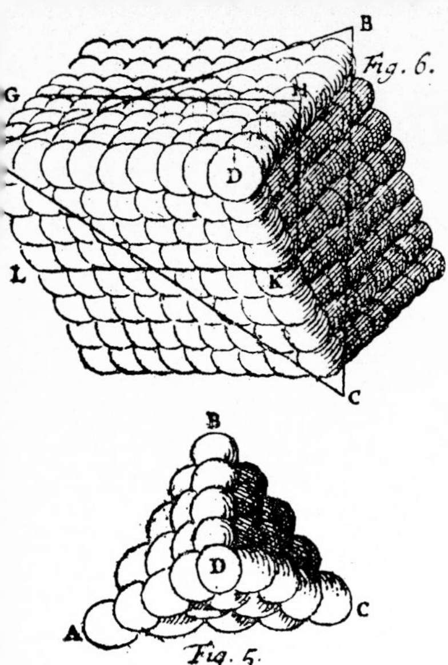

FIG. 83. Composition of a crystal of Iceland Spar according to Huygens, from *Tractatus de Lumine*

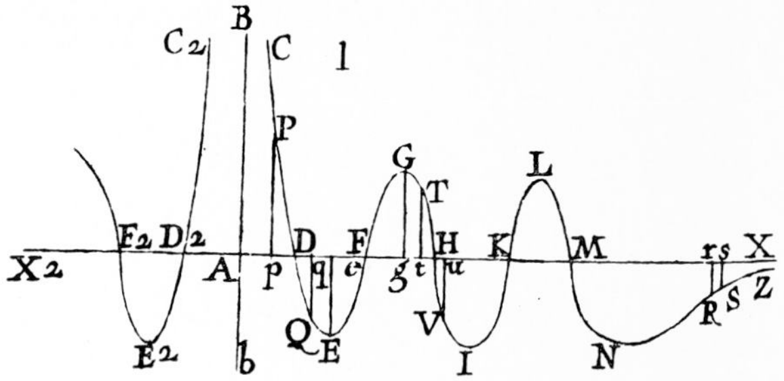

FIG. 85. Diagram showing the action between physical points according to Boscovich, from *Dissertationes*

IG. 84. Reproduction of an
iginal drawing by Huygens
owing double refraction
rough two crystals of
eland Spar

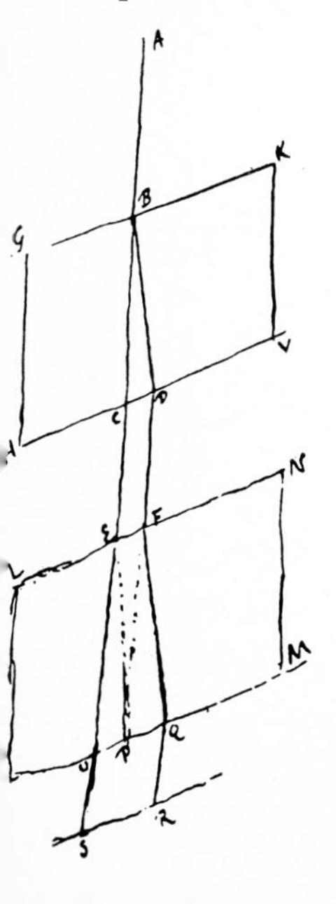

FIG. 86. Definition of 'quantity of action', from *Recherches des Lois du Mouvement* by Maupertuis

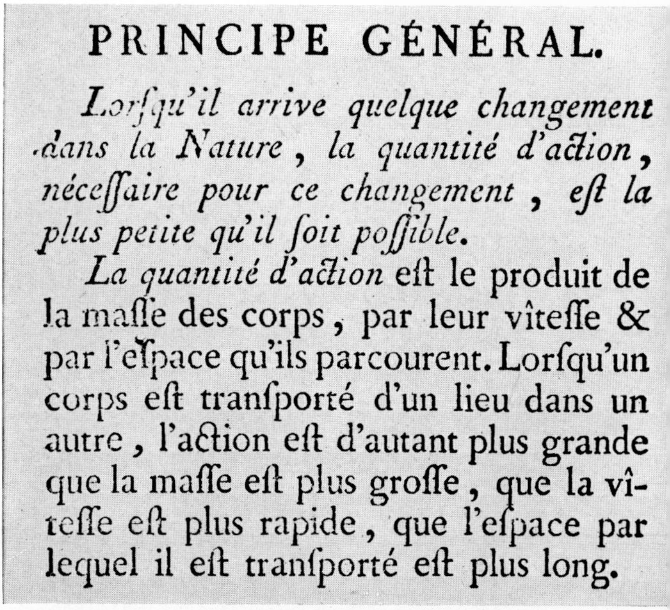

PRINCIPE GÉNÉRAL.

Lorſqu'il arrive quelque changement dans la Nature, la quantité d'action, néceſſaire pour ce changement, eſt la plus petite qu'il ſoit poſſible.

La quantité d'action eſt le produit de la maſſe des corps, par leur vîteſſe & par l'eſpace qu'ils parcourent. Lorſqu'un corps eſt tranſporté d'un lieu dans un autre, l'action eſt d'autant plus grande que la maſſe eſt plus groſſe, que la vîteſſe eſt plus rapide, que l'eſpace par lequel il eſt tranſporté eſt plus long.

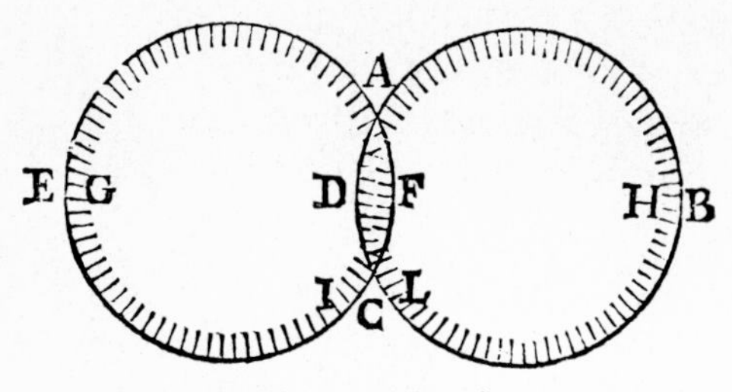

FIG. 87. Diagram showing two beams of sunlight overlapping. Grimaldi noticed that where they overlapped the area was darker, from *De Lumine*

FIG. 88. Thomas Young (1773–1829)

forming opaque bodies. But Huygens excluded the possibility that there could be '*mollities*', that is the absence of hardness of the particles, because silver and mercury were opaque and yet were among the best reflectors. As a more plausible solution Huygens suggested that the opaque bodies were composed of both hard and soft particles mixed together (unfortunately for his theory silver and mercury are in fact simple elements). The hard particles were responsible for propagation and also for reflection, while the soft particles dampened the motion and hence were responsible for the opacity of the body. Transparent bodies were those which did not possess these soft, damping particles. Having stated this he said no more, but seemed to think that he had said enough to allow him to pass immediately to refraction. This he explained by means of the envelope of secondary wavelets, assuming that the velocity of light changed when the medium in which it was propagated was changed. This explanation is so well known that there is no need to dwell on it. Huygens then extended it to total reflection, which took place when the envelope could not be formed in the medium where light had a greater velocity, and he explained why this type of reflection was more efficient (figs. 78 and 79). The delicate question of partial reflection from transparent sides was a hard test for this theory. Huygens attributed the cause of it to the reflection by the particles of the external medium which touched the reflecting surface and perhaps also to the reflection by particles of air and of other substances which were mingled with ether and greater than ether. He admitted, however, that he was not satisfied with this explanation,[34] especially since reflection also took place when the body was in a vacuum. How much mystery there was in this phenomenon which to all appearances appeared so simple! It was a stumbling block to scientists for many centuries, including men of great mental power such as Grimaldi, Newton, and Huygens.

The chapter on refraction ended with a simple and elegant proof of the principle of the shortest optical path, which Hero had proved in the case of reflection, and which Fermat, only a few years earlier, had extended to the case of refraction. Huygens then went on to

[34] Op. Cit. p. 32

study the double refraction produced by Iceland spar, and he gave an explanation which is a masterpiece.[35] Since there were two refracted waves, he concluded that with the Iceland spar there must also have been two velocities of propagation. Since one of the refracted waves followed the ordinary laws and the other did not, he concluded that the first had to be spherical while the other had some other form. Huygens assumed that the shape of this was an ellipsoid; he applied to it the principle of the envelope of secondary wavelets, and he deduced the behaviour for the various conditions of incidence. His results showed a remarkable agreement with experimental data.

There still remained to be explained the two velocities of propagation within the crystal. To explain this Huygens made use of the third mechanism of propagation in matter expressed earlier when discussing the phenomenon of refraction. One wave was propagated in the ether which was between the material particles, while the other wave was propagated in the particles themselves. If it was accepted that these particles had particular shapes, as was seen from the macroscopic structure of the crystal, it was not difficult to accept that the velocity had to be different in the various directions, so as to give to the wave the form of an ellipsoid (fig. 81). Huygens was extremely attracted by this idea. He tried to deduce important consequences from it, and to verify them by means of experiments, finding an agreement which he himself called 'marvellous'.[36] The experiments he carried out in the end created certain difficulties, and Newton made a point of calling attention to them. As we have already mentioned when discussing Newton's 'Query 25', when two rays, the ordinary and the extraordinary emerge from a crystal of Iceland spar and are made to pass through a second crystal, in general both of them are divided

[35] In 1690 Leibnitz wrote to Huygens: 'Mais quand j'ai vu que la supposition des ondes spheroïdales vous sert avec la même facilité à resoudre les phenomènes de la refraction disdiaclastique du crystal d'Islande, j'ai passé de l'estime à l'admiration.'
Oeuvres complètes de Chr. Huygens, 1901, Correspondence Vol. IX No. 2628 p. 522

[36] Op. Cit. p. 64

into two rays. For certain orientations of the crystal, however, the rays do not undergo any division, although when a ray of light is allowed to fall directly on the second crystal in this particular position, the doubling of the ray undoubtedly takes place. This surely means that the waves emerging from the first crystal are no longer identical with the usual waves, called 'natural', which entered the crystal, but must have an attribute of their own which makes their behaviour different from that of ordinary waves in relation to the second crystal. Huygens confessed that he was unable to define the mechanical reason for this phenomenon. He too returned to the idea of 'disposition' and wrote:

> . . . it seems that one is obliged to conclude that the waves of light, after having passed through the first crystal, acquire a certain form or disposition in virtue of which, when meeting the texture of the second crystal, in certain positions, they can move the two different kinds of matter which serve for the two species of refraction; and when meeting the second crystal in another position are able to move only one of these kinds of matter. But to tell how this occurs, I have hitherto found nothing which satisfies me. Leaving, then, to others this research. . . .[37]

And with this ends the part of *De Lumine* which contains Huygens' ideas on the nature of light. Huygens did not deal with many questions leaving many untouched, but those he studied he did thoroughly. From what he said explicitly, and from what we can gather in reading between the lines, he intended to complete existing theories and to fill any gaps in them. Probably he entered blind alleys because he always seemed to be in search of new models and new ideas to guide the progress of the waves. He never asked himself: what were these waves? By what laws were they ruled? What were the parameters on which they depended. Here again the lack of mathematical preparation reappears. We see Huygens who was driven by the evidence to postulate a 'form' or a 'disposition' in the waves, and yet who did not ask himself how could these

[37] Op. Cit. pp. 68 and 69
See also: Great Books of the Western World, Vol. 34. C. Huygens, *Treatise on Light* translated by S. P. Thompson, p. 601

waves be represented, and how could their 'form' be studied. But ideas were evolving and ripening.

It is interesting to note that the first studies on waves from a mathematical and quantitative point of view, are to be found in Newton's main work, *Philosophiae naturalis principia mathematica*. So Newton himself, with his genius and his work, was preparing the way for the triumph of that theory which he himself had opposed so explicitly.

CHAPTER SIX

The transition from the eighteenth to the nineteenth century

THE eighteenth century is generally accepted as the century of the triumph of the corpuscular theory of Newton. In our opinion this general belief leads to an overevaluation of the real development of facts, and as the distance in time increases the details tend to become obscure. The careful examination of texts and of the opinions of the time give us a very different idea of the state of knowledge. We must here differentiate between the public, even if informed, and the very restricted circle of specialists on the subject. The public does not look for details, it accepts only the main outlines with no means of criticising them, absorbs them, acclaims them as long as the ideas are easily understood and clearly expressed. It tends to have a mental inertia, and once it has accepted an idea it takes a very long time before it is able to cast it aside. It seems to receive the information from various sources, and tends to absorb more the philosophical than the technical content of ideas. It is not surprising, therefore, if in such an atmosphere, Newton's theory was widely acclaimed. During the seventeenth century, ideas had undergone an evolution due to Descartes and Galileo; the public asked for no more than material mechanisms which could take the place of the endless discussions of the Peripatetics.

To all those who were looking at the problem from the outside, in a general and summary way, it must have seemed a dazzling success to have 'explained' light with great simplicity by means of a model capable of explaining reflection and refraction within the framework of gravitational phenomena, and to have included in this also the explanation of colours (upon which Peripatetic philosophy still held strong sway). Even if there were some reservations and some obscure points, the general beauty of the construction was bound to suggest that there must have been at least some truth in the theories, and that the progress achieved by further studies would solve few mysteries which still existed and would finally incorporate them in the general framework. In addition it was believed that perfection did not exist in this world, and that new phenomena to be explained would always arise. Hence the fact that there were some phenomena which could not be understood at the time, did not necessarily invalidate the truth of the theory.

We must not underestimate the fact that diffraction and interference, and especially the first of these, are unfamiliar and little known phenomena. Even today, there are probably more people who know nothing about them, than there are people who have a thorough knowledge of them. In most text books on optics we are likely to find a long and detailed exposition of the general laws of geometrical optics followed by a much shorter chapter which explains that experimentally the subject is much more complex. If we make a detailed study, we find that the two phenomena of diffraction and interference affect considerably the concepts of geometrical optics. Nevertheless, there is a tendency to believe that diffraction and interference are negligible difficulties, and that both the phenomena and the laws regulating them, can be ignored without necessarily lowering the level of our general knowledge. If this happens today when we know the smallest details of these phenomena, it is easy to appreciate what it must have been like two centuries ago when these phenomena were insoluble mysteries, even for men of the mental stature of Newton and Huygens.

We know only too well that physical theories are temporary ideas which try to explain phenomena. Even when they are accepted, it is always possible to discover weak points in them, unless we are completely blinded by our preconceived ideas. If we do not wish to spend our lives as sceptics, always repeating that everything is temporary and that there are no absolute truths, we must be prepared to select and accept what appears to be good in a theory, and trust that the future will give us the explanation of the questions still unexplained. Indeed the public is very ready to accept this. In this situation, the accepting of one theory rather than another, becomes purely a question of taste or of opinion. Since a perfect theory does not exist, then it is a question of personal choice as to which theory we are prepared to accept from among the imperfect, and hence untrue, theories which are available. Perhaps it is a question of choice dictated by reasons outside the theory itself. Often it depends on the emergence of an active supporter of these theories, someone whom we like to call an 'apostle'.

Let us examine the work of Descartes, Grimaldi, Newton and Huygens. Descartes devoted himself to the study of so many questions that he gave very little time to the study of light, and he did not leave us a great contribution. Grimaldi worked very diligently on the question of light, but he never took a definite line; in any case he died young, even before the publication of his work. Huygens seems to have dealt with the question of light almost as if in his spare time, touching upon only some of the points of the great question, and giving us the impression that his attempt did not quite succeed. He concentrated, above all, on a very difficult and very little-known question, the phenomenon of double refraction, neglecting completely the subject of the theory of colours which, at the time, was very topical. After many years of study Huygens published his work only when he felt that his life was approaching its end, not wishing that it should be completely lost. Newton, on the other hand, the great genius, the world-renowned scientist, and the discoverer of the principles of mechanics, expressed and supported his ideas about optics for more than sixty years, with all his power, and with a tenacity and firmness that have become famous. He sustained all polemics without ever giving any sign of weakness or doubt, demonstrating instead an absolute conviction and a total faith in the triumph of his ideas. At the same time, he enriched experimental physics with a mass of new data and results. The public was bound to be attracted by the ideas of such a great man. They were easily understood, conclusive, attractive, and in addition there were no rival ideas which were so complete and so easily understood. Furthermore, Newton's ideas came well within the framework of the philosophical ideas of the period. Were there any gaps and unexplained facts? Perhaps, but Newton took good care not to stress too much the weaker points of his theory, and whenever he found himself in a tight corner, he charged his successors with the task of solving any difficulty. Thus the public, who in reality could not appreciate the importance of the unsolved points because they were too difficult to grasp, did not lose confidence in the future, and quietly accepted the fact that the theories were bound to follow slowly their natural development at the hands of the successors of the great man, now old and tired. It is not surprising that, for a century, the corpuscular theory had a great success among the general public; indeed, it would have been very strange if it had not.

Among the specialists in the subject, the situation was very different. When we examine the books which deal with the question of light in the period which followed Newton, we ask ourselves whether in fact a real corpuscular theory was ever accepted by the more advanced scientists of the time. They knew that the form that Newton had given to his work on optics was not purely a rhetorical form. As we have had the opportunity of mentioning before, Newton, like Grimaldi, never pronounced himself decisively in favour of a corpuscular theory because he felt that he was unable to support it. He knew very well the existence of the phenomena of interference, of diffraction, and of double refraction, and he also knew only too well that the corpuscular model could not explain these phenomena. His contemporaries had stressed this point very strongly; not only Huygens but above all Hooke, who was only seven years older than Newton and who was the secretary of the Royal Society. Relations between Newton and Hooke were very strained and often their discussions degenerated into strong polemics, so much so that Newton delayed the publication of his treatise on optics until 1704, one year after the death of Hooke.[1] Another great opponent of Newton was Leibnitz (1646–1716). While Hooke, who was a great experimenter in optical phenomena, attacked Newton mainly on experimental and technical grounds, Leibnitz criticized him from a philosophical point of view, particularly his tendency to coin new words, such as 'force' and 'attraction' which only appeared to be more significant, in order to explain properties which were just as mysterious and obscure as those postulated by the Scholastics. In this Leibnitz saw a betrayal of the mechanistic concepts of the natural philosophy of the century.

Among the more notable opponents of Newton in the middle of the eighteenth century we must mention Euler (1707–1783), a famous mathematician who was a scholar of great authority. He was

[1] See T. Young, *Lectures on natural philosophy* London, 1845, Lecture XL, p. 378

known, among other things, for having proved that refraction and dispersion were not proportional, and that it was possible to correct the chromatic aberration of objectives by a suitable combination of two lenses made of different glass. In *Dioptricae* we find the following 'Scholium':

> This defect due to the different nature of rays which appeared so serious even to the great Newton that he thought it was impossible to eliminate it in refracting instruments, can almost certainly be easily eliminated now, at least as far as the coloured edge of the images is concerned, to which Newton specially referred; so that, at least for this effect there is no longer any need to have recourse to reflecting telescopes.[2]

Euler could express himself thus in 1769 because he had studied this question mathematically fifteen years earlier and also because John Dollond (1706–1761), in England, had already made achromatic objectives, thereby bringing about a real revolution in the design of optical instruments. This happened in 1757, almost ninety years after Newton had completed his study of colours, when he had asserted his famous 'error'.

Neither the theoretical demonstration of Euler nor the practical work of Dollond seemed to impress the physicists of the time, perhaps, it has been suggested, because they considered this to be more of a technical argument than a fact of great scientific importance. We, however, believe that this idea did not receive the attention it deserved because it was striking at the roots of an idea which was already dead. Another serious difficulty for the corpuscular theory was caused by the fact that dispersion was not proportional to refraction, that is to say that the variation in the attraction of rays of different colours was not a function characteristic of the rays themselves, but was a complicated function of the quality (and not only of the mass) of the attracting matter. But this argument was rather too subtle for the general public which was unable to assess the importance of it. For specialists in optics this argument was not necessary as they already had plenty of arguments to enable them to evaluate the inconsistency of the corpuscular theory, and this new argument did not add very much that was new.

[2] See Euler, *Dioptricae* 1769, Part I, Chap. VI, p. 265

Huygens, Hooke, Leibnitz, and Euler represented the open opposition against the corpuscular theory, but the supporters and followers of this theory were themselves not less dangerous. It is symptomatic that the corpuscular theory never progressed beyond the stage where Newton left it. Among its supporters no one made any appreciable contribution or solved any of the problems bequeathed by Newton. The reason for this is obvious.

Let us now consider the work of an important scientist who declared himself a follower of Newton. We are referring to Father Ruggero Giuseppe Boscovich of Ragusa (1711–1787) who was a well-known if now forgotten astronomer. When Newton died, Boscovich was only fifteen years old and his life spanned almost the whole of the eighteenth century. Of great interest is a booklet that he published under the title of *De Lumine*, which consisted of two *Dissertationes*. The first dealt with the general laws of optics as they had been expressed by others, particularly by Newton; the second contained a theory of the author himself, which was supposed to represent the completion of the theory of the great master. Although Boscovich began by stating that optical theory had made great progress, he admitted that there was still 'a long and difficult way to go', and that nature seemed to have hidden its secrets. He added:

> In fact even if we are completely sure about the movements and the directions of the rays, there are still some things which have not yet been investigated enough and discovered; even among those things which at first sight appear absolutely certain and obvious and which are accepted as a solid foundation for all the others, there are some, if we reason properly and reject preconceived ideas, that are deeply rooted and to which the mind is strongly attached by the continual habit of asserting them, which cannot be considered as having been proved, or even justified by honest reasoning.

Boscovich did not argue against the 'causes of motion', or the 'forces', nor the 'inherent constitution of the various rays', but he attacked the rectilinear propagation of light and proved that light did not travel in a straight line! With common sense and acumen he called attention to the fact that most of the time those who asserted that light travelled in a straight line followed a vicious circle, because they simply called rectilinear the path followed by light. If, on the other hand, we wished to give an absolute meaning to this geometrical concept, then those who asserted that the trajectory of light was rectilinear could be compared to those who judge that the surface of the Earth was flat from the form of the surface of a small pool of water. If we took into account the motion of bodies, the inhomogeneity of transparent media, and the diffraction or inflection of light, we could never say that light followed a trajectory which was strictly rectilinear. Here we have the example of a follower and supporter of the corpuscular theory attacking one of the fundamental arguments which was both in favour of the corpuscular theory and against the wave theory of light! Nevertheless he admitted that experiments, within useful limits, supported the theory of the rectilinear propagation of light, indeed upon this idea are based exact sciences such as surveying and astronomy. '*Hoc ipso successu, elati animi.*' If we are sincere we must admit that the rectilinear propagation of light is not proved, it is only accepted until the contrary can be proved.[3] Having thus established this hypothesis progress can be made. It can, however, only be accepted if we accept the corpuscular theory. Paragraph 33 of the first Dissertation explains this very clearly. Light could not consist of pressure because pressure in fluids did not follow a rectilinear propagation. It could not consist of waves, because these when they passed through a hole spread in all directions just as a sound coming through a window can be heard in the whole room. We are left only with the emission of corpuscles: '*Superest, ut in effluvio quodam consistat corpusculorum*...'. This shows that the corpuscular theory was accepted only because all the others were excluded. The Dissertation continues by resuming Newton's optics, indeed even by quoting in several places the original text, concerning reflection, refraction, dispersion, and colours. One particular sentence is very interesting:

[3] P. Boscovich, *Dissertatio De Lumine*, De Rossi, Rome 1749, p. 17, paragraph 30

> There are other properties on which depend the colours seen in thin films, in soap bubbles as well as the colours of natural bodies, which are much more difficult to explain and which physicists very often neglect.[4]

The theory of 'fits' is then given with the original words used by Newton. Boscovich then commented: '. . . but on this matter we ought to increase our knowledge'. Even the sentences, in which Newton left to posterity the thirty-one Queries, are given by quoting the original words of Newton, and this is followed by the comment of Boscovich:

> These are his words, but after him, as far as we know, no one has progressed in this investigation beyond the limits reached by him.

The first Dissertation ends with a passing remark on the complexity of the double refraction produced by the Iceland spar.

Boscovich passed to the second Dissertation with the aim of going beyond the limits reached by Newton. But his work demolished Newton's ideas rather than strengthening them with new arguments. Perhaps in order to give some value to his own contribution, Boscovich called attention to the weak points of Newton's ideas. He did this bluntly and in some cases even with irony. For example, when dealing with the 'fits of easy transmission or easy reflection', he quoted word by word Proposition 12 of Part II of Book II of *Opticks*. He gave the passage where Newton renounced the search for a theory relating to the nature of these 'fits' after having mentioned those waves which ought to precede the corpuscles, and prepare them for 'easy transmission or easy reflection'. Finally he called attention to the fact that Newton himself, in formulating Query 29, where he repeated the same ideas about 'fits', expressed himself with a little more confidence when he said:

> Nothing more is requisite for putting the rays of light into fits of easy reflexion and easy transmission, than that they be small bodies which by their attractive powers, or some other force, stir up vibrations in what they act upon. . . .

[4] Op. Cit. Paragraph 89, p. 41

Boscovich added: 'But we absolutely believe that something else is 'requisite'.[5] He made a detailed criticism of the weak points of the corpuscular theory without touching on 'Newton's error', diffraction and double refraction, because he wished to introduce the idea that matter consisted of physical points endowed with a sphere of action. This action was alternately attracting or repelling as a function of the distance between two material points and followed a periodic law. In order to represent this law he gave us a diagram where the abscissae represented the distances between the two physical points and the ordinates represented the forces; these were repelling when the ordinates were positive and attracting when the ordinates were negative (fig. 85). With this model of matter he explained everything:

> With a curve of this form, we say that we can explain very well all the general mechanical properties of bodies and many of their particular properties; so we think that all must derive from it. From this we can easily determine the form and the immutability of the basic elements of matter, we can explain mobility, impenetrability and extension of bodies, the relation between action and reaction upon which depends the collision of bodies, the reciprocal action of particles of matter at very small distances that are so numerous and so varied, and gravity that is directly proportional to the larger masses on which it acts and inversely proportional to the square of the distances, the cohesion of the parts, the rigidity and fluidity, the elasticity, hardness, density, rarefaction and six hundred others; and among these the properties of light this being the main purpose of this dissertation.[6]

We must recognize that the ideas of Boscovich were far from modest. At the same time we must admit that in comparison with the ideas of Huygens, according to which matter consisted of many atoms, spherical in shape and tangent to each other, the concept comes very near to the more modern ideas of a highly rarefied matter. As far as light was concerned, Boscovich attempted to show that his hypothesis based on attraction and repulsion, which

[5] Op. Cit. Paragraph 90

[6] Op. Cit. Dissertation II, Paragraph 7

naturally also applied to luminous particles, enabled him to overcome the greater part of the difficulties which stopped Newton. On the other hand Boscovich did not accept either Fermat's Principle of Least Time or Leibnitz's Principle of Least Effort. He only considered rectilinear propagation, transparency, opacity, emission, absorption, reflection, refraction, and dispersion. He reached paragraph 129 after 57 pages of dissertation, and he found that from his model 'naturally follows that which concerns diffraction'[7] but he considered that there was no reason to dwell on it.[8] There was also the question of the Iceland spar; but this too, according to him, could easily be explained by his theory. The dissertation, however, was already too long and it was time to stop. This was a great pity.

Boscovich, in a few pages, gave us a picture of the frame of mind of the supporters of the corpuscular theory. Newton had left to posterity ideas which were fated to be criticized by his opponents and to be destroyed by his supporters. These ideas could only be favourably received by superficial or non-theoretical minds, or by those physicists who, in general, were silent on the delicate points and omitted any discussion of difficult phenomena. This attitude was mostly found in schools where ideas were formed which lasted a life-time, thereby preventing the great majority of people from criticizing or even following the progress of ideas. This explains why for nearly a century the views of the opponents of the corpuscular theory were ignored and often laughed at; and by the end of the eighteenth century, even the best among the learned men were completely dominated by Newtonian ideas. Indeed they were even more convinced than Newton, and had ended by accepting as great discoveries some of the ideas which Newton himself was not prepared to assert, but that he had only put forward not to leave too many awkward gaps in his theories. In spite of all this the corpuscular theory was dead. It had died a natural death because in such an extremely favourable century and climate of thought, it remained stationary 'within those limits already reached by Newton'.

[7] Op. Cit. Dissertation II, p. 57

[8] Op. Cit. Dissertation II, p. 58

Theories are valuable only if they are fertile, if they are not they die.

For those interested in the function of dogmatism in science, it is worth noting that the Newtonian theories of the eighteenth century constituted a real and true dogma. The revolutionaries of the seventeenth century had struggled against the dogmatism of the Peripatetics to establish a rational method and a clear critical mentality in the scientific field. The result has been obvious. The dogmatism of the Peripatetics was supplanted by the dogmatism of the Newtonians which, unfortunately, was even stricter and more domineering than its predecessor. The Peripatetic dogma was clearly stated and was used as a method, while the Newtonian was not called dogma but was considered to be the result of experimental and rational investigations and, as such, was considered to be unquestionable. In practice it was forbidden to criticise it or to lay any stress on phenomena which did not come within its general framework. From this it appears as if the scientific world is destined to pass from one dogma to another, just as if dogmatism were an absolute necessity.

In the theoretical field, after the tentative work of Boscovich, nothing very remarkable occurred in the eighteenth century concerning the nature of light. The experimental field, however, was enriched by the discovery of new phenomena. We must recall the discovery of phosphorence (as it is called today) made by Vincenzo Cascariolo, who was a shoemaker in Bologna. While experimenting with secret chemical processes, he tried to calcine a fossil which was found near the top of Mount Peterno, in the neighbourhood of Bologna. By chance he noticed that this stone emitted a faint glow when taken into the dark after being exposed to light. According to Priestley, from whom we take this information, the discovery was made about 1630.[9] Throughout the seventeenth century, scientists payed little attention to this reported phenomenon, and we only find some passing mention of it in the writings of Kircher

[9] J. Priestley, *The History and Present State of Discoveries relating to Vision, Light and Colour*, Johnson, London 1772.

According to others the discovery was made in 1603 or 1604.

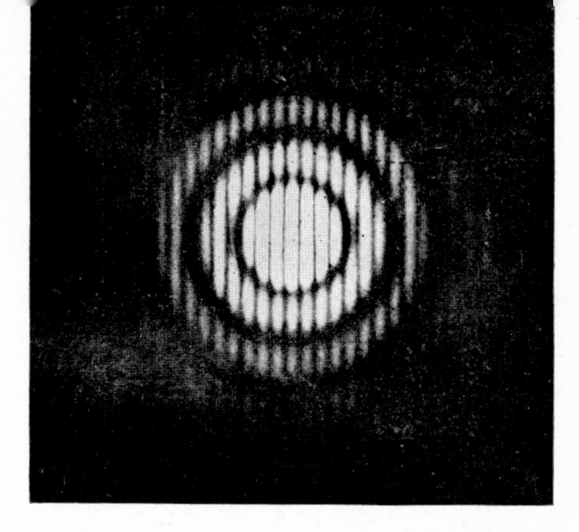

FIG. 89. The appearance of interference fringes in the experiment described by Young

FIG. 90. Augustin Fresnel (1788–1827)

FIG. 91. Diffraction pattern produced by an obstacle with parallel edges illuminated by a point source, from *Lezioni di ottica ondulatoria* by V. Ronchi

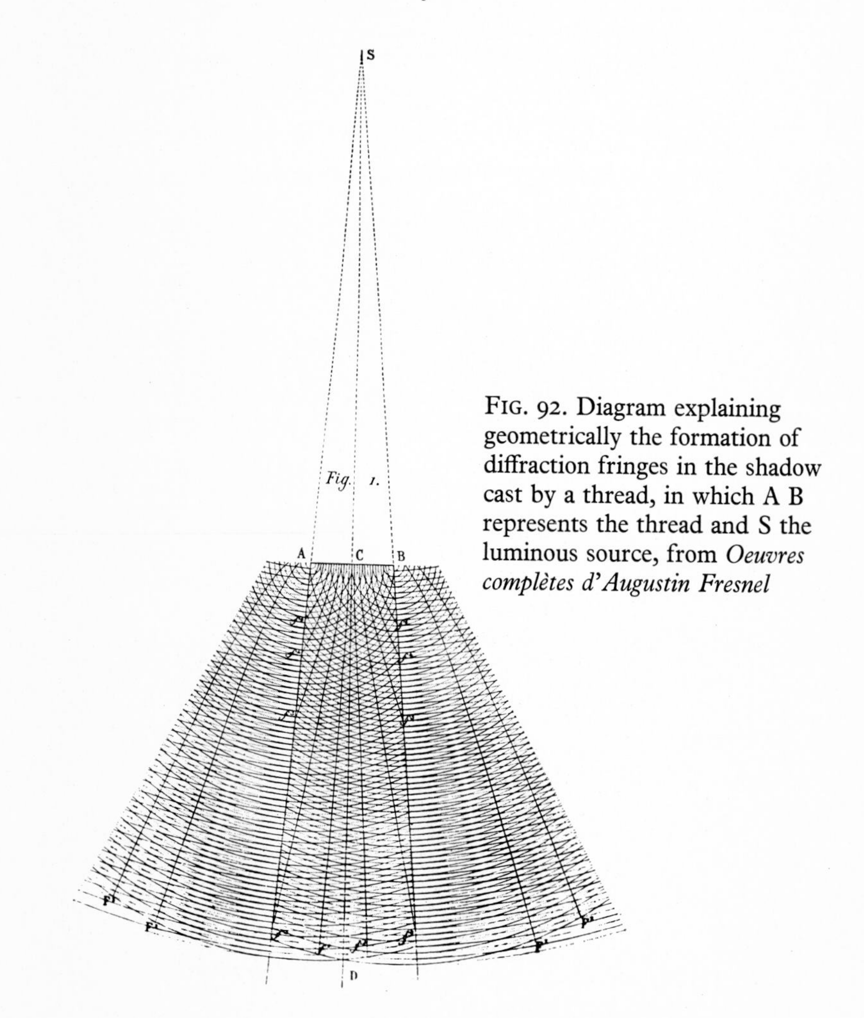

FIG. 92. Diagram explaining geometrically the formation of diffraction fringes in the shadow cast by a thread, in which A B represents the thread and S the luminous source, from *Oeuvres complètes d'Augustin Fresnel*

FIG. 93. Diagram illustrating Fresnel's theory of 'effective rays', from *Oeuvres complètes d'Augustin Fresnel*

FIG. 94. Type of obstacle used by Fresnel to test experimentally the theory of 'effective rays', from *Oeuvres complètes d'Augustin Fresnel*

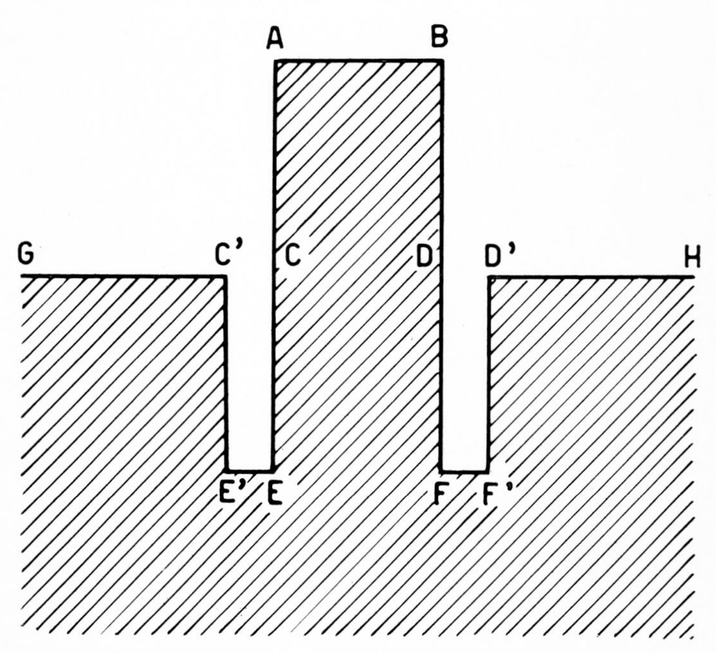

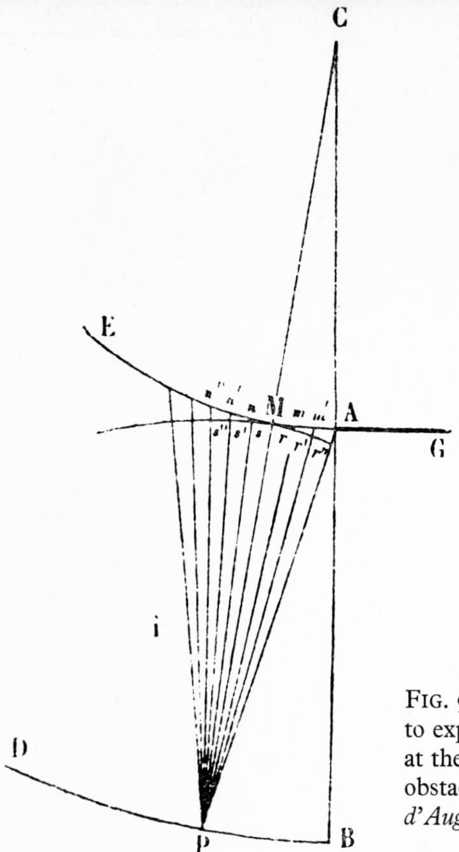

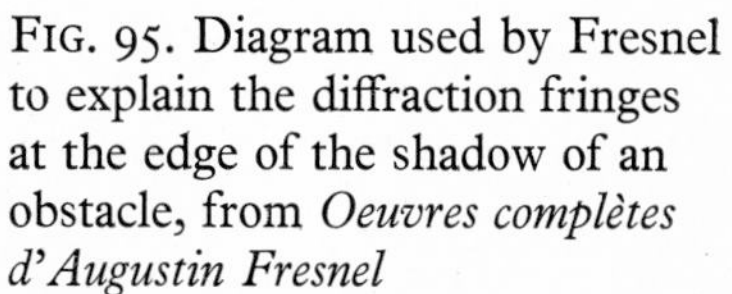

FIG. 95. Diagram used by Fresnel to explain the diffraction fringes at the edge of the shadow of an obstacle, from *Oeuvres complètes d'Augustin Fresnel*

FIG. 96. Concentration of light at the centre of the shadow produced by an opaque, circular screen illuminated by a point source, from *Lezioni di ottica ondulatoria* by V. Ronchi

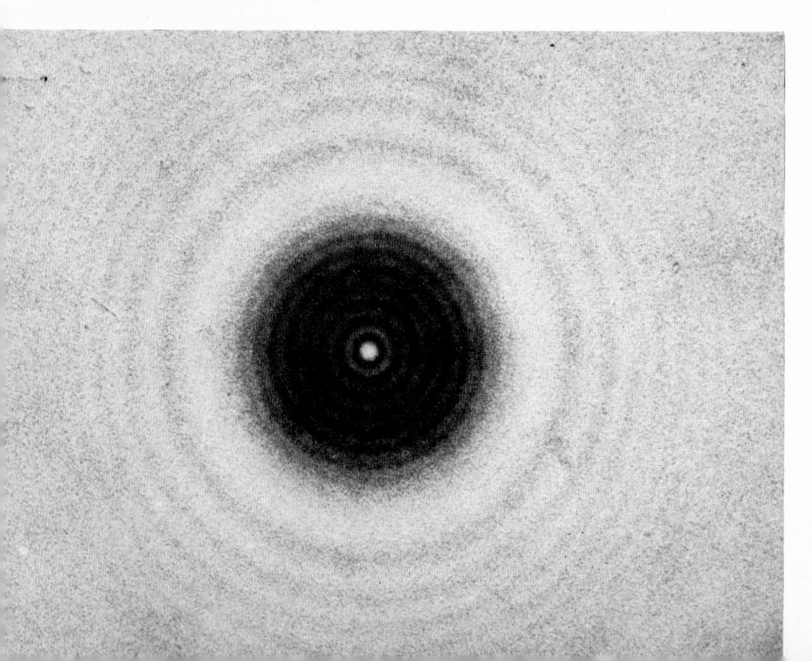

(1601–1680). When, however, towards the end of the century, interest in theories of the nature of light increased, many investigations were carried out on this strange phenomenon in the hope of finding some evidence which could either confirm or destroy existing theories. Many experiments made by Count Marsili of Bologna, by M. Lemery, Galeazzi, Zanotti, Baldwin, Father Beccaria, Du Fay, and by many other investigators, led to the discovery of many other natural and artificial substances, which showed the strange property attributed to the 'phosphorus of Bologna', as the fossil of Mount Peterno was called. In particular, Du Fay discovered in the course of his investigations in this field that a diamond, after being heated for some time, emitted light when in the dark. This phenomenon was considered by the majority of scientists to favour the corpuscular theory, although, as experiments increased, it showed such a variety of manifestations and such strange behaviour from the chromatic and thermal point of view, that it neither confirmed nor demolished the theories on the nature of light.

Another discovery, which at the time seemed to bring some support to the corpuscular theory, was that of the 'aberration of light' made by Bradley (1692–1762), the third Astronomer Royal. Since December 20th, 1725, he had noted anomalies in the positions of some stars observed with his zenith telescope at Greenwich. At first he attributed these anomalies to errors of observation, but later he noticed that they were regular and were related to the time of the year when the observations were made. Bradley explained this phenomenon as being the combination of the motion of light from the star with the orbital motion of the Earth around the Sun. Even this did not help to give a better definition of the nature of light.

The eighteenth century, however, saw the beginning of very important studies and measurements concerning the measurement of light. Generally speaking the question of photometry was discussed in most of the early books on optics which we have mentioned. Lucretius in *De Rerum Natura* discussed the dazzle produced by staring at a very bright source such as the Sun. He also discussed how the visibility of illuminated objects improved when the observer was in darkness, while dark objects could not be seen so well by someone who was in bright light. More developed ideas on these questions are found in Alhazen's book. We have already mentioned such questions as the study of the persistence of vision, and, in particular, how this phenomenon was used to strengthen the view of the existence of light as a physical agent. There are also discussions on subjects which today we would call visual sensitometry.

The ideas on photometry which are found in Leonardo's manuscripts are much more advanced. Many of these ideas were in due course collected in the *Trattato della Pittura*. As is well known, Leonardo introduced the technique of shading in painting, which in technical terms is nothing more than the photometric gradation of penumbrae. He made many interesting studies of shadows and of penumbrae. As an example we can quote his Codice C Folio 22r:

> But if the brighter source of light is distant from the body casting the shadow and the less bright source of light is near, it is certain that the shadows can be made to appear of equal darkness and brightness. If between two sources of light a solid body is interposed at an equal distance from both, the body will cast two shadows in opposite directions, and the darkness of the two shadows will depend upon the power of the two sources of light which produce such shadows.

In the *Trattato della Pittura* we find many photometric references. For example, in paragraph 672 it is stated:

> If a great light of low power is equal to a small light of great power . . . [and in paragraph 626:] *Which body assumes greater quantity of light.* A greater quantity of light will be assumed by that body which by greater *lumen* is illuminated.

In these passages, and in many others, we find not only visual considerations but we begin to detect the outline of problems of a purely photometric character, with quantitative evaluations, even if they are not expressed by numbers. It is extremely interesting to find that the question of the equality of shadows produced by light sources of different power is already being mentioned.

In the books on optics of the sixteenth and seventeenth centuries we find a considerable increase in the interest in photometric questions, particularly from the quantitative point of view. For instance, it was observed that the illumination of a surface diminished when the distance between the surface and the source of light increased and when the inclination of the rays increased. Already we have quoted one of the fundamental hypotheses which is found in *Photismi de lumine et umbra* by Maurolico:

> The denser rays illuminate more intensely; rays of equal density illuminate equally.

There is a particularly interesting drawing that acts as heading to Book IV of *Opticorum Libri VI* by the Jesuit Father Aguilonius, published in 1613 in Antwerp. Book IV is devoted to studies of photometry of a classic type. The strange thing is that the drawing mentioned above (which is purely ornamental), shows people observing a screen illuminated by two sources of light, which is remarkably similar to one of the early types of photometers.

It is not until the eighteenth century that we find a real development in quantitative measurements in photometry. Some attempts in this direction were made by Buffon who tried to measure the percentage of light reflected by a mirror. It was Bouguer (1698–1758), however, who really tackled the problem. With great care and ability, he succeeded in surmounting the many difficulties which were encountered in this type of measurement. His first *Essai d'Optique* was very successful; so he set himself the task of writing a more complete work, to which he dedicated most of his life. He died in 1758 and his work was published later in 1760, edited by De la Caille, under the title *Traité d'Optique*. Because of his work, Bouguer was considered to be the scientist who, after Newton, had obtained the most important results in the study of the properties of light. He laid the foundations of photometric technique by highlighting some of the most difficult points, such as that of the null method. This consisted of carrying out measurements by varying, in a known manner, one of the quantities to be measured, so as to make it equal to the other quantity. Thus all the observer had to do was to decide when the two quantities appeared equal and was not expected to make an evaluation of small differences.

By these methods, carried out with extreme care, Bouguer succeeded in measuring the amount of light reflected from bright and from diffusing surfaces, as well as measuring the transmission of light. Bouguer's work is of philosophical importance to our study. In the eighteenth century, a slow but progressive transformation took place which tended to detach scientists from the models of the past, and to introduce false prejudices which were generally accepted as evident and unquestionable. We have already shown in Chapter III, that, in the scientific world before the eighteenth century, it was generally believed that an external agent existed, emanating from the Sun and from flames, which was capable of exciting light when it struck the eyes of an observer. There was also a widely-held belief that light existed only in the 'soul' or, that is to say, in the mind of the observer. There existed a similar distinction, though less explicitly stated, in the case of colour. To minds less used to abstract thought, it appeared evident that colour was 'something' located on the surface of coloured bodies, and this was the 'something' which pure light, the white light of the Sun, acquired when it illuminated the bodies and which was carried to the eyes of the observer. As we have seen, it was one of the most important achievements of Grimaldi to have proved that colour was not an entity separate from light.

We have already mentioned that scientific texts written in latin, which was then the official international language, distinguished between the external physical agent and the internal psychic effect, by using two different words, the first agent was called *lumen* and the second *lux* (see page 61). It was generally said that the Sun 'illuminated' bodies, but there did not exist a corresponding word with *lux* as its root, since such a word would have been meaningless. This distinction, which was so clear and explicit, was generally accepted even in the seventeenth century. Galileo quoted a comparison which left no doubts on his point of view on this question. Speaking of light he said that if he brushed the tip of his nose very gently with his finger, he felt a tickle, but this did not mean that he could conclude that the tickle was in his finger. Both Newton and Huygens had no doubts on this score. On the contrary, we can say that the reason why the corpuscular and wave theories of *lumen* arose was due to the belief that *lumen* was not *lux*, but could be anything, which by acting upon the retina produced the vision of *lux*. Only with this assumption was it possible to conceive *lumen* as consisting of a swarm of very small particles or as a train of waves. Had this not been the general belief, or, in other words, had people been convinced that *lux* was an external agent and the same thing as *lumen*, then the model used to explain *lumen* would have had to take into account both light and colour. No one could imagine how it was possible to talk about waves that were light and dark, or red and yellow.

Newton in Book I, Part II of his *Opticks*, expressed himself very clearly on the subject of colours. Let us quote him again:

> The homogeneal light and rays which appear red, or rather make objects appear so, I call rubrific or red-making: those which make objects appear yellow, green, blue, and violet, I call yellow-making, green-making, blue-making, violet-making, and so of the rest. . . .

He insisted on this point, explaining that in the rays 'there is nothing else than a certain power and disposition to stir up a sensation of this or that colour', just as the vibration of a bell or of a musical string enables us to hear sound. Gradually these definitions tended to be forgotten. If we consider very carefully Newton's words we notice the beginning of a change. When he wrote the first sentence, he stated that he called the rays 'rubrific' or 'red-making'. With the first word he wanted to convey that the rays were not red but allowed the observer to see a red colour, and this would have been very clear. But he added that he also called them red-making which he justified with the following words:

> And if at any time I speak of light and rays as coloured or endued with colours, I would be understood to speak not philosophically and properly, but grossly, and accordingly to such conceptions as vulgar people in seeing all these experiments would be apt to frame. For the rays, to speak properly, are not coloured. . . .

After this explanation in the original text he no longer mentioned the word 'rubrific', but called the rays red; later, when dealing with rays of other colours, he simply called them yellow, green, blue, and violet, dropping the suffix '-making'. It is not surprising, therefore, if a century later, physicists had forgotten his initial reservations and accepted that rays were actually red, yellow, green, blue, or violet. A world which was at the time bent on isolating and localizing everything to the exterior, and which was trying hard to forget that there existed an observer with a mind and a personality, was only too happy to follow on the slippery path on which Newton himself had ventured. Light followed a similar fate to colour. Attention became fixed on *lumen*, and *lux* was forgotten. When latin ceased to be the official international language of science, modern languages used only one word to denote both *lumen* and *lux*. The word which already existed in these modern languages was that in common use to denote the opposite to darkness, namely the 'light' that one sees. In correct usage this word stood for what was known as *lux* in the scientific language of the seventeenth century, but physicists adopted it to mean *lumen*. This greatly contributed to the idea, even in cultured circles, that there was no real reason to distinguish between *lux* and *lumen*, and that light was something objective, physical, and external to the observer. It was a false concept which introduced serious errors and endless ambiguities, but which is deeply rooted even today.

This mistaken concept was already widely accepted when Bouguer laid the foundations of experimental photometry, and this was bound to lead to the problem of how to measure light. Photometry was born to solve this problem. At the beginning its development was very slow, because the investigators were faced with the inevitable consequences of the fundamental ambiguity on which their work was based. They were convinced that they were carrying out physical measurements, while in reality they were carrying out psychological experiments. Unknowingly Bouguer indicated methods that gave the illusion of obtaining objective results, which thus strengthened the illusion that it was a question of carrying out physical experiments. Among these we can recall, in particular, the 'null method', which we have already mentioned. We must admit, however, that Bouguer's measurements did have very marked characteristics of physical experiments.

Photometry was launched on this path, and in the following century there were investigators who were more and more convinced that it was the right path.

We must now say something about the vicissitudes of Fermat's Principle, which were due mainly to the complacent mentality of the eighteenth century.

In Chapter IV we have given a detailed account of Fermat's Principle. We can add that Newton, and even more his followers, banished the dangerous idea advanced by Fermat. The followers of Newton supported the idea that the velocity of light was greater in dense substances than in rarefied ones. Supporting this idea was not an easy task, and it was made even more difficult by Fermat's reasoning.

In 1682 Leibnitz published a note in the *Acta Euroditorum* in Leipzig, in which he exposed the way by which he could reconcile the law of 'economy in nature' with the existence of the greater velocity of light in denser media. He remarked that the first law was not to be understood to mean that light followed the shortest path nor that it travelled a path which required the least time, rather that 'light follows the easiest path'. By the 'easiest path', he meant the one for which the distance travelled, multiplied by the resistance encountered by the light in the medium, was a minimum. Starting from this definition, and using the calculus which Leibnitz had founded in competition with Newton, he obtained the well-known law, concerning the ratio of the sines in the case of refraction. In order to avoid reaching the same conclusions as Fermat, Leibnitz made the assumption that the velocity of propagation was proportional to the resistance encountered by the light in the medium in which it travelled. He justified this incredible assumption by an analogy: the narrower the bed of a river, the faster is the flow of water. Naturally this whole theory was soon forgotten.

Both Clairaut and Mayran gave some thought to the same question. Mayran in 1723, in a memoir presented to the Académie des Sciences in Paris, related the history of the polemics between

Descartes and Fermat. He called attention to the

> embarrassment and the impotence met so far when attempting to make the law of refraction agree with the metaphysical principle of the maximum economy of nature.

In a memoir presented to the Académie des Sciences in 1744, Maupertuis (1698–1759) touched upon this very delicate point of the theory of light. He believed that he had found 'the agreement of different laws of nature, which until now had appeared incompatible'; this was in fact the title of the memoir. The laws to which he referred were: the maximum economy of nature, the constancy of the ratio of the sines in the case of refraction, and the greater velocity of light in denser media. Maupertuis was not acquainted with Leibnitz's memoir on the same subject, published in 1682, and he related the history of Fermat's Principle to show that the statement was unclear as to whether it was a question of the path being the shortest or the time being the least. Naturally he decided that it could not be the path that was the shortest, because this did not agree with the experimental law of refraction. Neither could it be a question of the least time, since this would of necessity require that the velocity of light should be less in the denser media; but he added that such a metaphysical law should not be too narrow. He said:

> Pondering on this question I thought that light when it passes from one medium to another, thereby already leaving the shortest path which is the rectilinear path, could also not follow the path of least time. Why should time have preference over space? Light not being able to go at the same time by the shortest path and by the least time, why should it go by one of these paths rather than by the other? Therefore if it does not follow either of the two it takes a path which has a much more real advantage, the path which it follows is that for which the 'quantity of action' is a minimum.[10]

Maupertuis explained at once what he meant by 'quantity of action'. A more complete and precise exposition of the 'general principle' is given in a memoir with the title *Recherches des lois du mouvement*, presented to the Berlin Academy two years later.

[10] *Oeuvres de Maupertuis* Alyon 1768, Vol. IV, p. 16

> When some change takes place in nature, the quantity of action necessary for the change is the smallest possible. . . . The quantity of action is the product obtained by multiplying the mass of the bodies by their velocity and by the distance travelled. When a body is carried from a point to another, the action is all the greater as the mass is greater, as the velocity is greater and as the distance the body travels is longer.[11]

Maupertuis applied the principle of the minimum quantity of action to refraction, in this case ignoring mass, which has no importance in a question of a single body. Simply by means of calculus he obtained the law of the constant ratio of sines when the velocity was greater in the medium in which the angle between the ray and the normal was the smaller, that is in the denser medium. Here we have the solution of the enigma: nature is very economical and Newton's ideas are in agreement with the laws of nature. We must admire the ability, finesse, and tenacity with which these scholars have struggled to break through the barriers surrounding this mystery. At the same time it is perplexing to see the flexibility of the so-called 'exact sciences' which by cast-iron laws of logic and by the infallible help of mathematics can lead to conclusions which are diametrically opposite to one another.

Maupertuis's reasoning, which seemed to him and to his followers a masterpiece of genius and rectitude, is completely forgotten today, and this spares its author from being accused of having made an error. The kindness of posterity is not in fact very generous, because it robs Maupertuis of the honour of having defined the concept of action, a concept which is widely used today. In fact, Plank's constant h, which is the 'quantum of action', is the product obtained by multiplying mass, velocity, and distance, exactly as Maupertuis had defined it. (Fig. 86.)

We have described all these sterile and inconclusive debates to show the complacent and mistaken mentality of the eighteenth century concerning light. Now we have reached the beginning of the nineteenth century and our history can record another development of special interest.

[11] Op. Cit. p. 36

CHAPTER SEVEN

The triumph of the wave theory

THE views of the supporters of the corpuscular theory at the beginning of the nineteenth century are very well described by Malus (1775–1812). In his memoir *Théorie de la double refraction de la lumière dans les substances cristallisées*, which received the prize of the Académie des Sciences in Paris in January 1810, he began with these words:

> Optical phenomena have the special advantage of being able to be measured with great precision and of being linked by a small number of mathematical laws. These laws are independent of the hypotheses which can be advanced concerning the nature of light, because, whether we agree with Newton that it consists of a very rarefied fluid emitted by all the parts of luminous bodies, or whether we agree with Huygens that it is produced by vibrations of an ethereal fluid, the path of the rays is always the same. The hypothesis of Huygens is subject to great difficulties; it appears incompatible with the chemical phenomena produced by light; the theory of emission is more credible and agrees better with our physical knowledge. In this work, therefore, I will adopt Newton's opinion, not as an unquestionable truth, but as a starting point and in order to interpret the operations of the analysis. It is a simple hypothesis which in reality has no influence on the results of the calculations.

The theory of emission could not have been treated in a more humiliating way. It was more than a century old and had no dangerous rivals, yet was treated here not by an opponent but by a supporter, simply as an expedient and 'not as an unquestionable truth'.

Malus had undertaken the study of double refraction in order to compete for the prize which the Académie des Sciences had offered in 1808 for the following subject: 'To give a mathematical theory supported by experiment, of the double refraction which light undergoes in travelling through various crystals'. The very fact that such a subject had been suggested by the Académie proves that it was not satisfied with the existing ideas on this phenomenon. Malus won the prize in 1810, and published his memoir. In the memoir what is even more important than the theory of double refraction, is the discovery which is still known today as Malus's experiment. He showed that, to prevent a given beam of light dividing into two when passing through a crystal of Iceland spar, it was not necessary, as it had been thought, for the beam to pass first through a double refracting crystal; it was enough that the beam should have undergone reflection. Malus expressed himself very clearly in his memoir:

> On the basis of the experiments which I have described, the characteristic which distinguishes the direct light [natural light] from that which has been subjected to the action of a crystal, is simply this: that the former can always be divided into two beams, while in the latter this faculty depends upon the angle formed by the principal sections of the two crystals.[1] [A few pages later we find the following:] When a beam of light travels through a diaphanous substance, part of the rays is reflected by the refracting surface and part by the surface of emergence. The cause of this partial reflection has so far eluded the physicists. Only rarely has attention been paid to this question and no attempt has been made to study the modifications of this strange phenomenon, which seems to be an exception to the general laws of refraction and which is as strange as that of the doubling of images. We shall see however, that there exists, between these two types of phenomena, a very close link and that in some way they may be considered to depend upon the same cause, since they present identical effects. We shall begin by the simplest example and by the easiest experiment. If we let a beam of light fall upon a surface of still water at an angle of 52° 54′ to the vertical, the reflected light has all the characteristics of one of the beams produced by the double refraction of a crystal.[2]

This is Malus's discovery. As a follower of the emission theory, he raised again a question which all the physicists of the eighteenth century had prudently left alone, and at the same time he created unexpected complications. Partial reflection was not understood, and it appeared that reflected light was modified in a way similar to light which had undergone double refraction. Malus concluded:

> In general in all phenomena of this type, the plane of reflection replaces the plane of the principal section of the first crystal.

[1] E. L. Malus, *Théorie de la double réfraction de la lumière dans les substances cristallisées'* Baudouin, Paris 1810 paragraph 48, page 219

[2] Op. Cit. paragraph 49, page 221

This takes place when the reflecting substance is diaphanous; the question is much more complicated when the reflection is produced by polished metals. Malus succeeded in showing that the emission theory could explain these phenomena, if the idea hinted at by Newton in Query 26 was accepted. This was that corpuscles of light had a certain innate asymmetry which, generally, in a swarm of particles of natural light, had a random orientation, but that, in passing through a double refracting crystal, underwent an alignment and became ordered. By analogy with magnetic bodies Malus suggested that the corpuscles of light had poles, so he called the light which had particles all uniformly oriented, 'polarized light'. This term is still used today. He found that there was no difficulty in assuming a polarizing action in reflection, because the reflecting surface had the power of orienting the corpuscular bipoles. At the same time he thought that his discovery was even more difficult to reconcile with Huygens's theory than all the other phenomena already known at the time.[3] He added:

> In fact, if the simple phenomena of reflection are different for the same angle of incidence, something which could not happen in the case of a wave system, we must conclude not only that light is a substance subject to the forces which act upon other bodies, but also that the shape and the disposition of its molecules have a great influence upon the phenomena.

Malus asserted this after recalling that Huygens himself had recognized that his hypothesis could not explain the phenomena produced by light which had undergone a double refraction, when it was made to pass through a second crystal. Only too often did Huygens' opponents reproach him for this admission. Newton had learnt from the experience, so he was always extremely careful not to stress too much the weak points of his constructions, in spite of the fact that he had much more reason than Huygens to do so. Perhaps it would have been better if Huygens had left to posterity the task of explaining this mystery rather than denying the possibility of explaining it. Posterity indeed explained the mystery left by Huygens but not those bequeathed by Newton.

[3] Op. Cit. paragraph 54, page 239

The corpuscular theory was approaching its end, indeed its life had been so long only because of the absence of another theory. The studies and discussions of the eighteenth century had caused the ideas to ripen. Euler had taken up the wave theory, and he had added the concept already put forward by Grimaldi, that colours were due to the wavelengths of the waves. Euler had also studied the propagation of waves in diaphanous bodies. Although his ideas did not withstand criticism because they were too materialistic, they had the merit of stimulating discussion, and of preparing the ground for his successors. Real progress was brought about by two young men of genius, neither of whom belonged to the academic world. One was an English physician, Thomas Young, and the other a French government civil engineer, Augustin Fresnel. Thanks to the extraordinary ideas of these two intruders, the corpuscular theory had become, within a few decades, of historical interest only, and the wave theory acquired fundamental importance in the physics of the nineteenth century.

The title of Proposition XXII in *De Lumine* by Grimaldi is rather strange:

> Light can sometimes by its intervention make darker a surface which had been previously illuminated by another source.[4]

The phenomenon is well known today and can be expressed more or less by the words used by Grimaldi; but the experiment described by him does not give what today is known as the phenomenon of 'interference of light'. In his experiment two cones of light, obtained by letting the light of the Sun enter a darkened room through two small holes, were observed on a white screen placed at such a distance from the holes that the two circular sections partly overlapped (fig. 87). In these, and in similar conditions such as might be created by letting light enter through any two openings and then by making the two beams partly overlap by means of a mirror, Grimaldi noted that the edge of the zone where the two beams overlapped was less bright than the surface

[4] Grimaldi, *De Lumine*, 1665, Propositio XXII, page 187

illuminated by only one of the two beams.

In these circumstances, because of the manner in which the experiment was carried out and because of the account he gave of the observation, it is impossible that the phenomenon of interference, in the ordinary sense of the word, was what Grimaldi really observed. In fact, in order to see the interference in the conditions described by Grimaldi, it would have been necessary to have a source of light with a diameter of the order of a few seconds of arc, rather than the Sun which has a diameter of 32 minutes of arc. Probably the effect he described was partly due to diffraction and partly to the contrast of the various illuminated zones. It is true that Grimaldi ended his Proposition by saying that '. . . this depends on the skill of the experimenter and on the keenness of the eye of the observer'.[5] It is strange that a hundred and forty years later, a device very similar to that described by Grimaldi was used by Young to show that light added to light can darken a screen instead of further illuminating it. Thomas Young (1773–1829) was a rebel against dogmatism of any sort and a lover of new and unexpected ideas. He was barely thirty years old when, probably in the course of his medical studies, be became interested in the questions of vision and the nature of light. He was far from satisfied with the explanation of Newton's rings by the theory of 'fits', especially since a century and a half of criticism by both followers and opponents of the theory, had done nothing to make the theory progress, but had, on the contrary, shown its deficiencies. It was obvious to him that some new theory was needed to explain this phenomenon. (Fig. 88.)

When confronted by a thin film of varying thickness which showed alternate light and dark zones, Newton thought this was the effect of alternate reflection and transmission, and he must have been encouraged to observe that the transmission fringes were complementary to the reflection fringes. He understood that the second surface was necessary; to the first surface he assigned the task of exciting the periodicity of the 'fits' of the corpuscles of light and to the second the task of analyzing them. Thus he reached the

[5] Op. Cit. page 190

ideas expressed in Proposition 12 of Part II of Book II of *Opticks*. The partial reflection of diaphanous surfaces, which seemed to undermine the theory of light, compelled him to modify his model of 'fits', with the result that the function of the first surface of the thin film could not be understood. The easiest approach was to ignore the whole question. Young had no intention of ignoring it, and with great insight he considered as essential for the formation of the fringes, the partial reflection of light upon both the first and second surfaces. He thought that the light reflected by the second surface was added to that reflected by the first surface. The fact that there were black zones where there was no light reflected by the first surface, led him to the conclusion that in these places, the two lights must cancel each other, while in the light zones the two lights must strengthen each other. Newton had not mentioned the fact that, according to his theory of 'fits', the fringes could not be black when seen either by transmission or by reflection, because in each case the light reflected by the first surface of the thin film had always to be present.[6]

Young, in a note published in the *Philosophical Transactions* of 1802 under the title 'An account of some cases of the Production of Colours, not hitherto described', stated that he had discovered a law which could explain these phenomena until then 'insulated and unintelligible'. These are his words:

> The law is, that wherever two portions of the same light arrive at the eye by different routes, either exactly or very nearly in the same direction, the light becomes most intense when the difference of the routes is any multiple of a certain length, and least intense in the intermediate state of the interfering portions; and this length is different for light of different colours.[7]

In 1804, again in the *Philosophical Transactions* under the title 'Experiments and calculations relative to physical Optics', Young

[6] This observation however, was made by Fresnel in his: *Complément au premier memoire sur la diffraction* in *Oeuvres Complètes* 1846, page 52

[7] Thomas Young, *Philosophical Transactions* 1802, page 387

related various experiments on diffraction; by comparing them with those performed by Newton he concluded:

> From the experiments and calculations which have been premised, we may be allowed to infer, that homogeneous light, at certain equal distances in the direction of its motion, is possessed of opposite qualities, capable of neutralising or destroying each other, and of extinguishing the light, where they happen to be united; that these qualities succeed each other alternately in successive concentric superficies, at distances which are constant for the same light, passing through the same medium.[8]

He then remarked that by means of this law the fringes and the colours of thin films could easily be explained. Since Newton had already noticed that if water was substituted for air in a thin film the rings became narrower, Young concluded that the velocity of propagation of light had to decrease in denser media because he assumed that the periodicity 'was bound to be unchanged in a given quantity of light'. Sound behaved in a similar manner, and therefore Young concluded: 'there must exist some close similarity between the nature of sound and that of light'. How many times in twenty-three centuries has it been repeated that there was a close analogy between the nature of sound and that of light! This is still being said but it is not a convincing argument.

In his thesis for his degree, Young had discussed the question of the production of the human voice, and this enabled him to obtain a deep knowledge of the laws relating to sound and to the propagation of waves. He was well acquainted with the works of both Newton and Boscovich, and the latter must have made a deep impression on him because Boscovich attempted to demolish the preconceived idea of the rectilinear propagation of light and hinted that diffraction had to be considered. Young studied closely the emission theory and strongly criticised it as far as diffraction was concerned. He remarked what Grimaldi had already demonstrated and what Newton must have known, that the diffraction fringes were independent of the acuteness of the diffracting edge and of the substance of the edge itself. This circumstance had been ignored

[8] *Philosophical Transactions. The Bakerian Lecture* 1804, page 11

not only by Newton but by all those who insisted on speaking of the 'inflection' of light. Young took up again this question and attempted a new experiment. He produced bright fringes in the dark shadow of a thread and then showed that all these fringes disappeared if the light, which grazed only one of the two edges of the thread, was intercepted by means of an opaque screen. This proved to him that to form the fringes, it was necessary that light should come from both sides of the thread. This experiment showed at last that the idea of the rectilinear propagation of light, on which the theory of emission was based, was false, since diffraction showed that light was capable of going around obstacles.

All these considerations,[9] and many others which we have already mentioned concerning transparency and the constancy of the velocity of light, led Young to favour the wave theory. He took up again the ideas advanced by Huygens and used them to explain the fundamental phenomena of optics. He completed his work with an experiment that became famous and which still bears his name. He made the light from a point source pass through two extremely small holes close to each other, just as Grimaldi had done with the light of the Sun. The two holes became two sources of coherent light, and because of diffraction produced diverging beams. Where these beams overlapped, interference fringes were formed which obeyed the law that Young had formulated and which he had called the 'law of interference'. He carried out measurements and deduced the periodicity of the light for various colours, thus he obtained 1/36,000 of an inch (0·7 μ) for red and 1/60,000 of an inch (0·42 μ) for violet (fig. 89). All this important work was done over a period of only three years. Inevitably, side by side with all these wonderful ideas, there were some errors. Young believed so much in his law of interference that he tried to use it to explain diffraction. Although it may be plausible to con-

[9] Laplace (1749–1827) who was a follower of the theory of emission, had shown that a star 250 times greater than the Sun would remain invisible, because light would not be able to emerge. Also stars slightly smaller than this would emit light with velocities which depended on their masses.

sider the bright fringes in the shadow behind a thread as simple phenomena of interference, produced by the overlapping of two beams of light, it is not so easy to explain the dark fringes which are found in the illuminated parts around the shadow, which had been described by Grimaldi, Newton and others. Young believed that here was a case of interference between the direct light and the light reflected by the surface of the diffracting edge. He was wrong. Grimaldi had already proved that the reflected light had nothing to do with it.

Although the ideas of Young were indeed very important, it was unfortunate that his manner of expressing them was neither systematic, nor clear, nor precise, and as a consequence he was strongly criticized by his contemporaries. Young did not adopt the right method to gain the approval of public opinion. He alienated it by the vehemence of his youth, and consequently he was not taken seriously. Lord Brougham's public attack in the *Edinburgh Review* is well known and Young was stunned. He wrote an article which was not accepted by any periodical, and in the end he published it himself in the form of a pamphlet. The serious accusation levelled at him was that he had been too critical of Newton's ideas and experiments and that he had exposed 'presumed' errors, which certainly a man like Newton could never have made. Newton always had to be right. Young in his reply maintained that he had never doubted Newton's greatness and that he had always admired his great genius and his great discoveries; he added: 'But much as I venerate the name of Newton, I am not therefore obliged to believe that he was infallible'. Therefore he did not consider it wrong to call attention to those points where Newton's reasoning seemed to be at fault and to advance a theory which could explain easily and completely so many of the phenomena which all the followers of Newton had deliberately neglected. This reply of Young is reproduced in *Miscellaneous Works of the late Thomas Young* published in 1855 in London by John Murray. It ends with the following note by George Peacock, the editor:

> Of the preceding most masterly reply, which was published in the form of pamphlet, it was stated by its author, that, *one copy only was sold*;

> it consequently produced no effect in vindicating his scientific character, or in turning the current public opinion in favour of his theory.[10]

Such was the level reached by Newtonian dogmatism!

However, Young's ideas had already been launched. While those of Newton, although deeply revered, were beginning to die after a century and a half of sterility, those of Young, although derided, were to bear wonderful fruit after ten years. This was mainly due to the work of Augustin Fresnel (1788–1827), a government civil engineer (fig. 90).

Fresnel, having completed his studies at the *Ecole Polytechnique* of Paris, was employed in the south of France on road construction while he was in his early twenties; he spent about ten years there. In this very practical environment he did not lose his interest, probably acquired while a student in Paris, in scientific questions of an advanced nature. As a relief from his daily routine, he used to give some thought to theories about the nature of light. Often in his ledgers, side by side with the entry of workmen's pay, there appeared notes and remarks about light and similar problems. In the end he decided to write a memoir which he called *Rêveries*, and sent it to Ampère. We do not know what there was in this work and he himself forgot completely the content of it. From his behaviour in the years which followed we can assume that he was not very impressed by the Newtonian theory. In those years of scientific isolation, he formed the view that light could not consist of material corpuscles and that it was necessary to return to the concept of waves. It was purely a way of thinking, dictated by common sense, because Fresnel certainly was not in a position to go deeply into the question as he could have done had he been in an academic environment. His cultural background in this particular field of science was indeed meagre, as is shown by the following passage of a letter written to his brother Leonor in Paris, on May 15th, 1814. After asking him for a copy of *Physics* by Haüy, he added:

[10] *Miscellaneous Works of the late Thomas Young* edited by George Peacock, publisher J. Murray, London 1855, Vol. I, page 215

I would also like to have some literature that would bring me up to date with the discoveries made by French physicists on the question of polarization of light. I have seen in the *Moniteur* a few months ago that Biot had read a very interesting memoir on polarization of light at the *Institut*. I rack my brains in vain but I cannot guess what it is.

Less than a year later, in April 1815, political events changed the course of his life. Fresnel, who openly sided against Napoleon on his return from Elba, was dismissed from his employment and was detained by the police at Mathieu, near Caën. Finding himself in a small village free from the daily cares of his work, he was able to devote all his time to his 'dreams', and began to correspond with Arago (1786–1853) who was an astronomer at the Paris Observatory. It is interesting to record that although Arago was two years older than Fresnel, he was barely thirty years old. Although he had one of the most coveted posts and he was also a member of the Académie des Sciences, he had led an adventurous life and he could not have had a very wide culture nor many definitely formed ideas. This was to be an advantage for Fresnel. Young, who had stated important facts, was crushed by Lord Brougham; Fresnel who, on the other hand, had started timidly, found Arago who encouraged and supported him.

In July 1815, Fresnel wrote to Arago to ask his advice and directions on the question of the theory of light. Arago advised him to read Grimaldi, Newton, the treatise of Jordan, and the memoirs of Lord Brougham and Young. Fresnel, however, did not know any English nor, probably, any Latin, the use of which as a scientific language was, in any case, declining, so he did not read any of the books recommended. Instead he began to experiment, and on October 15th, 1815, he sent to the Académie des Sciences his first memoir entitled *La diffraction de la lumière*. The first word itself is a revelation, it is no more a question of inflection but of diffraction. The memoir begins with an attempt to demolish the emission theory. The first argument is a totally new one: light and heat must have the same nature because a black body, when illuminated, becomes warm, and a very hot body, on the other hand, emits light. According to him interplanetary space was occupied by heat, and if this consisted of particles how could it be possible for the light of the Sun to reach us with the usual velocity, if in its journey it was to meet so many milliards of thermal corpuscles? On the other hand, heat passed from air into a colder body, but then saturation and equilibrium were reached; why did this not happen in the case of light?

This first criticism levelled by Fresnel against the emission theory ended with some observations on vision, on the strangeness of the theory of 'fits', and on the polarity of the particles constituting light. It was a very doubtful criticism. Then Fresnel continued:

> It seems to me that the wave theory lends itself better to the explanation of the complex behaviour of the phenomena of light; and since at this stage there reappears the analogy with sound and the usual objection that waves go round obstacles, I have decided to study shadows.

Naturally he ended by observing the diffraction fringes, as Grimaldi and Newton had done. While these two were determined to see in this phenomenon, particles which deviated from the rectilinear path, Fresnel concluded from it that there were no particles. This was indeed a new situation.

Fresnel began his observations and measurements with rudimentary means, in fact he recounted that the micrometer used for his early observations had been made for him by the village blacksmith. In spite of this he saw and discovered many things that his predecessors had never seen. He used a special technique which can be considered as fundamental to all his work. In order to observe the diffraction fringes, it is necessary to have very narrow light sources, and hence the sources used are of very small intensity. At that time the light of the Sun had to be used, as that was the only bright source of light available. All Fresnel's predecessors, while carrying out this type of observation, made use of a hole in the shutter of a darkened room, as Grimaldi described, and allowed the beam of light to fall on a screen of white paper or on ground glass. They interposed between the source and the screen a diffracting object, such as holes, slits, edges, threads, and similar objects, and observed the effect of them on the screen. Fresnel, in order to see more clearly, made use of a magnifying glass. This led him to use ground glass instead of a paper screen to avoid obstructing the path of light with his head. He related that one day, while carrying out this observation, the ground glass which had not been properly fixed, slipped uncovering half of the field. With great surprise he noticed that in the half of the field uncovered by the glass, the fringes were more clearly visible than those on the ground glass. From then on in his observations he omitted the ground glass completely and received the beams of light directly on the magnifying glass and hence in the eye. This small incident, happening at the right moment, had important results. The great progress brought about by this apparently slight change in the experiment has only recently been fully appreciated. In this way Fresnel had available a much greater luminous flux, which enabled him to reduce the luminous source to very small dimensions. Instead of a hole in the wall of the room, he used a little drop of honey placed over a hole in a thin metallic strip; this acted as a little lens of extremely short focal length, giving a microscopic image of the Sun. Moreover, it enabled him to use an eyepiece with a filar micrometer which, even though made by a village blacksmith, allowed him to take measurements to a hundredth of a millimetre, something that no one had ever been able to do before. This new system of observation allowed him to detect how the fringes originated at the edges of diffracting obstacles, again something that no one had ever done before. This happened because Fresnel had put inflection on trial, as it were. He followed the process of its formation, and in doing this found that the external fringes originated all at the same time and actually at the edge of the obstacle.

In order to explain the development of Fresnel's studies to those readers not familiar with these phenomena, it is necessary to recall that the shadow of a thin thread, generally of a few millimetres in diameter, when illuminated by a very narrow source of light, is not a clearly defined rectangle, but has a rather complex structure, as is shown in figure 91. In the interior of the shadow are visible numerous fringes parallel to the edges and equidistant from each other, while outside the shadow there are other fringes also parallel to the edges, but with decreasing width and low contrast. Fresnel, therefore, observed patterns similar to that of figure 91 with a magnifying glass of high power, approaching closer until reaching the thread itself and he saw that the external fringes appeared to decrease in width and to end at the diffracting edge. This was an embarrassing result for the theory of inflection. If we look at figures 70 and 71 we can see that the Newtonian theory predicted very different behaviour in the neighbourhood of the diffracting edges. Fresnel then took measurements. In each plane he measured the distance of the external fringes from the edge of the geometrical shadow and he found that 'a given fringe is not propagated in a straight line but follows a hyperbola whose foci are the luminous point and one of the edges of the thread'. This language, which is very strange from a mathematical point of view, must be interpreted as referring to the phenomenon in a section which is at right angles to the thread. The measurements made with the primitive micrometer did not lead to such a clear cut conclusion, it was reached by Fresnel when he assumed the hyperbolic form, and he found that the difference between the values calculated and those obtained by his observations were smaller than the errors of observation, which, in any case, were less than seven per cent. Fresnel suggested the idea of the hyperbola because it corresponded to the interference between the direct spherical waves, namely those emitted by the luminous point, and the spherical waves emitted by the edge of the obstacle. Figure 92 shows clearly this concept. In the same way he explained the formation of the internal fringes, and obtained an excellent agreement between the measurements and the calculations. Furthermore, Fresnel proved that these fringes were formed by the light which came from the two sides of the thread, by covering one side with a small sheet of paper. There was, however, only one point of disagreement between the calculations and the observations. The calculations for the bright fringes gave the results that the observations had given for the dark ones. Fresnel temporarily explained this by assuming that the rays reflected by the edge of the obstacle lost half a wavelength in the process of reflection.

Convinced that the experiment was clearly in contrast with the corpuscular theory and was in agreement with the wave theory, he forwarded his note to the Académie des Sciences. A few days later he also wrote directly to Arago asking his opinion and outlining further developments that he intended to describe in a supplementary memoir. Arago, who together with Poinsot had been appointed by the Académie to examine Fresnel's memoir, wrote to him on November 8th, telling him that in the memoir he had found many interesting facts, although some of them had already been described by Young, and he added:

> . . . but what neither he nor anyone else before you has ever seen is that the external coloured bands do not *move* in a straight line, when the observer moves further from the opaque body. The results you have obtained on this subject seem to me extremely important; perhaps they will serve to prove the truth of the wave theory. . . .

Arago continued by promising him all his support against the objections that inevitably would be raised against a theory so opposite to the fashionable one. It is easy to imagine how elated Fresnel must have been in receiving these comments and these promises of support. His enthusiasm must have reached a high peak, and as a result he concentrated all his energies on these studies. All this was the result of a combination of fortunate circumstances and of errors. It is incredible how profitable these errors were. In fact Grimaldi and Newton had already observed that the external fringes did not move in a straight line, and had Arago and Fresnel read Grimaldi's *De Lumine*, they would have found in Proposition I (No. 17) that this fact had been clearly stated. It is strange that something so vital to the corpuscular theory should have been overlooked and forgotten. Fresnel came back to the same idea thinking that he had made a new discovery, and Arago was so impressed that he became at once a supporter of these new ideas.

The report to the Académie des Sciences by Arago and Poinsot was read at the meeting of March 25th, 1816. It praised Fresnel highly and proposed that the memoir should be published. Fresnel, however, was not satisfied. He was concerned by the exchange between the dark fringes of the theory and the bright ones of the experiment, and he was even more unhappy about the mechanism that he had suggested, because in this first memoir there was still a considerable Newtonian influence. In order to calculate the interference, Fresnel took into consideration the direct rays and the rays inflected by the edges of the diffracting obstacles. In this reasoning we find a mixture of rays and waves, of inflection and of diffraction. Once again we must remark that, had he read *De Lumine*, he would have found fully proved (Proposition I, No. 18), that in the case of diffraction, reflection from the surface of the diffracting obstacle was not involved. He did not know this and he continued to look for a conclusive experiment. If the obstacle was asymmetric, that is if it had one very thin edge and the other very thick, and if the external fringes were formed by the mechanism suggested, they should have been hardly perceptible on the side of the thin edge and much clearer at the other edge. Fresnel carried out a test with a razor, the cutting edge being much thinner than the back, and the experiment produced identical fringes on both sides. He concluded that his theory, even if approved and praised by the Académie des Sciences, was without foundation. On the other hand, the agreement between the mechanism of interference and the experiment was evident, apart from the interchange of the bright and dark fringes. It was necessary, therefore, to find another source for the interfering waves other than the material of the obstacle.

It is possible to reconstruct step by step the evolution of Fresnel's ideas, but we shall limit ourselves to a brief summary highlighting only the principal points. In a memoir presented to the Académie des Sciences on July 15th, 1816, barely four months after the publication of the report which praised his earlier memoir, Fresnel stated the theory of the 'effective rays'. He abandoned the question of the material of the obstacle, and following in the wake of Huygens, turned to the secondary wavelets. Indeed he went one step further than Huygens. While Huygens had used the envelope of secondary wavelets, Fresnel now knew that the waves could produce interference and that, therefore, it was necessary to take into account the 'phase difference'. This complicated considerably his reasoning, but did not deter him, and the goal at which he aimed was worth his efforts. The first step in this direction was his statement of the theory of the 'effective rays'.

In figure 93, A G represents the obstacle and A the diffracting edge. A, B, C, C', C'' are points of the luminous wave from which originate secondary wavelets travelling in all directions. At a point F in the shadow of the obstacle, all the wavelets arrive but with different phase. The arc C E is drawn by taking F as a centre and by making the radius equal to F A plus half a wavelength; FC' is drawn equal to FC, plus half a wavelength, then FC'' is drawn equal to FC', plus half a wavelength. Now the waves which arrive at F coming from points on the line CC' are cancelled by those arriving from points on the C'C'', because they have opposite phase, and the same is true for all those which are on the left of C''. Finally, the 'effective rays' which arrive at F all come from the arc CBA, and since A and C already send waves which have opposite phase, the centre of gravity of the 'effective rays' is approximately in B. This new method allowed Fresnel to reduce the difference between the bright fringes obtained by calculations and the dark fringes obtained by experiment. The difference had resulted from the adoption of an erroneous theory, because instead of taking point A, it was now necessary to take point B which was a little further away. However, the difference was not zero, it was only diminished by about half. Progress was still insufficient on this point, but it was considerable in the sense that diffraction was now independent of the material of the obstacle and depended on the wavelets. Fresnel tried to find by experiment a confirmation of his new theory. If it were correct, the appearance of the phenomenon behind the obstacle AG should have remained identical whether the section of the wave CC'C'' was active or not. He then constructed an obstacle like ABFF of figure 94 which in the upper part allowed access to all the lateral wavelets, while in the lower part it eliminated those further away from the diffracting edges. He illuminated this obstacle with the usual point source, and observed that the diffraction fringes behind the upper part ABCD were different from those that were behind the lower part CDFF. The experiment, therefore, did not confirm the theory of the 'effective rays', but it indicated, without any doubt, that the wavelets emitted by the points of the wave-front, even if distant from the diffracting edges, participated in the formation of all the fringes behind the obstacle. The final step was described in a sealed note deposited at the Académie des Sciences on April 20th, 1818. We shall return soon to the story of this note. Meanwhile, to conclude the development of Fresnel's ideas, let us look at figure 95. The vibration at point P is the resultant of all the wavelets originating from the infinite number of points of the wave front EMA. In order to determine the intensity of this vibration, it is necessary to calculate an integral. Fresnel began by using practical and numerical methods and obtained results that were in complete agreement with the experiment. Later mathematicians solved the same problem theoretically. Thus Fresnel, by combining the principle of wavelets with that of interference, stated the principle of integration of wavelets which completely explained all diffraction phenomena.

The full significance of this synthesis is admirably expressed by Schwerd in a sentence in the preface of his book entitled *Die Beugungserscheinungen*:

> the wave theory predicts the phenomena of diffraction as exactly as the theory of gravitation predicts the movements of celestial bodies. [Fresnel's principle was stated in these terms:] the vibrations of a luminous wave in each of its points are equal to the sum of all the elementary movements which would be sent there at the same instant by each little part of the wave, acting individually and considered in any of its earlier positions.

The above statement is found in two simultaneous notes: the first was in the short note which, as we have said, was presented sealed to the Académie des Sciences on April 20th, 1818, and the second in a more extensive and complete note presented anonymously to the Académie just before the end of July 1818, as an entry to the competition announced on March 17th, 1817.

Among the members of the Académie des Sciences, there were many followers of the Newtonian theory of which Biot and Poisson were the staunchest supporters. They were alarmed by the successive memoirs of Fresnel in 1815, 1816, and 1817, and above all they were frightened by the increasing number of phenomena that this enthusiastic engineer succeeded in including in the general framework of his theories. These were all phenomena that the supporters of Newton had been compelled to leave well alone in order not to have to confess their inability to explain them. The situation was reaching dangerous proportions, and the challenge had been openly made. A decisive reaction was required. The supporters of Newton believed that the time had come to retaliate, and made the Académie des Sciences organize a competition with a prize for the following subject:

> The phenomena of diffraction, discovered by Grimaldi and later studied by Hooke and Newton, have recently been the subject of research by numerous physicists, especially Young, Fresnel, Arago, Pouillet, Biot etc. The diffraction fringes that are formed and are propagated outside the shadow of bodies have been observed, as well as those that appear within the shadow itself, when the rays pass simultaneously from the two sides of a narrow body, and those fringes that are formed by reflection on surfaces of limited extent when the incident and reflected light passes very near to their edges. So far, however, the movement of the rays in the proximity of the said bodies where inflection takes place, has not been sufficiently studied. The nature of these movements offers today an aspect of diffraction which needs clarifying because it contains the secret of the physical manner by which rays are inflected and separated in various bands having different direction and intensity. This has led the Académie to offer a prize for an investigation on this subject to be studied in the following manner:
>
> 1. To determine by precise experiments all the effects of diffraction of luminous rays both direct and reflected, when they pass separately or simultaneously, near the extremities of one or more bodies having an extent either limited or indefinite, bearing in mind the distance between these bodies, as well as the distance of the source of light from which the rays emanate.
>
> 2. From these experiments to obtain by means of mathematical induction, the movements of the rays in their passage in the proximity of the bodies.

This was indeed a Newtonian theme expressed in Newtonian terms, but no supporter of the Newtonian theory came forward. Only Fresnel and someone else unknown, who was not even considered, entered for the competition. The committee which had been elected to judge the entries, and which consisted of Biot, Arago, Laplace, Gay-Lussac and Poisson, chose Arago as chairman, and this meant the triumph of Fresnel and the final downfall of the corpuscular theory. At the open meeting of the Académie des Sciences of 1819, the prize was presented to Fresnel, and this date marked the collapse of Newtonian ideas.

The struggle was over, but followers of old ideas do not disappear suddenly nor can they be converted quickly not even when faced with the most complete evidence. Many similar examples have already been quoted in previous pages. The Newtonian ideas had been so majestic and so important that traces of them could not be expected to disappear overnight, indeed even today some of these ideas are still evident in some form, and it is natural, therefore, that at the time of Fresnel they should still have some influence. A good example of this is found in the behaviour of Poisson. The memoir of Fresnel, which fills 135 pages of his *Oeuvres Complètes*, was a treatise on the wave theory of light, and was full of theory and experiments. In it was found the principle of integration of wavelets, as we have already mentioned, and this was followed by the statement that this led to a very complicated integral which was not always possible to calculate by means of the mathematics known to the author. In the report of Arago we find the following:

> One of the members of the committee, M. Poisson, had deduced from the integrals given by the author, the strange result that the centre of the shadow of an opaque circular screen should be illuminated as if the screen did not exist, when the rays fell with an incidence which was not very oblique. This consequence was tested by a direct experiment and the observation confirmed perfectly the calculations.

Figure 96 shows the phenomenon very clearly. Nevertheless, Poisson remained a staunch supporter of the Newtonian theory, and even after such a brilliant experimental proof he did not accept the new ideas. Until his death, which occurred twenty-two years later, he continued to speak of his '*filets de lumière*' of which he never gave a satisfactory definition. He remained a convinced adversary of Fresnel with whom he carried on a very lively argument. His tenacity and his attachment to the ideas of his youth did not succeed in resurrecting a concept which, in any case, had to be considered as dead. Biot also showed an incredible tenacity in maintaining the ideas of Newton. It is said that even as he was dying, nearly forty years after the complete downfall of the corpuscular theory, he was still attempting to explain by means of corpuscles all the enormous number of new phenomena which had been discovered with the help of the wave theory.

Having won his first great battle, Fresnel found opening before him a boundless vista, full of promises and possibilities, but also bristling with difficulties. The phenomena of polarization and of double refraction had now to be incorporated in the new theory, if he could succeed in this his triumph would be complete, otherwise his early success might be changed into defeat. He started to work with all his energy, but in better conditions than when he first began. In fact he had been reinstated in his original appointment and could go frequently to Paris, where he was a welcome guest at the laboratory of Arago. In 1819 he was transfered to Paris, to the department dealing with lighthouses, and it is a well-known fact that even in this field, so different from the one with which we are dealing, he made important contributions which today still bear his name. This move enabled him to be much closer to cultural centres and to have at his disposal experimental means more suitable to the delicate experiments dictated by the evolution of his ideas.

In reading the memoir that Fresnel presented to the Académie des Sciences in 1819, we are amazed by the mastery he displays of the wave theory, which was a new subject and little studied in those days, even in other fields of physics. Fresnel seems to use this theory with a skill and finesse equal to that of the most expert in optics of our times. One episode is of particular significance. In studying the fringes within the shadow of a thin thread Fresnel, as we have already stated, had thought of a test, which incidentally Young had already carried out. This test consisted of obstructing, by means of an opaque obstacle such as a sheet of paper, the passage of light from one side of the edge of the thread. In these conditions all the fringes within the shadow disappeared, not only the fringes which were on the side of the obstructed edge. This experiment was considered as a heavy blow to the theory of inflection. One day though, Arago very worried, informed Fresnel that the same result could be obtained by substituting a transparent plate of glass with plane parallel faces for the sheet of paper near the edge of the thread. This could cast doubts on the value of the proof against inflection. Fresnel, however, immediately solved the mystery by showing that when the plate of glass was thick and the light was not homogeneous, the fringes within the shadow became of such a high order that they were not visible, but when the plate of glass was extremely thin, the fringes could still be seen, although displaced laterally. Arago performed the experiment and was amazed; everything happened as Fresnel had predicted.

Young had carried out the classic experiment of interference by using two holes illuminated by a narrow source of light, but this experiment did not appear very convincing because, on the whole, light travelled by two separate routes, and the partisans of inflection could advance some ideas, even if very complicated, to explain the fringes which were formed. Young explained it by the principle of interference, but for the theoretical effect to be complete he had to prove that it was not possible to explain it by means of inflection; this was a very hard task. Fresnel conceived the well-known devices of 'mirrors' and of 'biprism' that bear his name today. In these conditions, interference still took place, but it was no longer possible to speak of inflection. Finding the interference fringes by means of Fresnel's mirrors is a very delicate operation when a monochromatic radiation is not available, and this is due to the fact that the position of one mirror with reference to the other, has to be determined to the nearest micron. Fresnel was aware of this and calculated everything with great certainty, in addition he succeeded in using this as an argument in favour of the wave theory. According to him if a theory allows the prediction and execution of such a delicate experiment where chance has no place, then this theory must correspond to reality. He made a very severe criticism of the theory of 'fits' used to explain the fringes in thin films. He proved that in some particular cases, for example when the film is inclined to the beam of light, the result of the experiment is just the opposite of the result predicted as a necessary consequence of Newton's theory. The wave theory, on the other hand, explained everything very naturally and with great simplicity. Having mastered the wave theory, Fresnel tackled the phenomena of polarization and of double refraction, and he obtained results which amazed the scientific world. These results survived the criticism of his contemporaries and of his successors, and are still accepted today.

It must not be thought, however, that his work was easy. It progressed rapidly because the productive period was about five years, from the time of his prize-winning memoir on diffraction. The ideas he expressed were so revolutionary that a time came when even his friend Arago thought it prudent to withdraw his support. By then, however, Fresnel could stand on his own feet, especially as the Académie des Sciences had elected him a member. In those five years, from 1820 to 1824, the contribution of Fresnel to the study of the phenomena of polarization was considerable. A complete analysis of his contribution would lead us too far from our subject; instead we shall limit ourselves to a summary of the arguments which interest us directly.

Fresnel, as we have said, had a complete grasp of the wave theory and of the behaviour of waves in the phenomenon of interference. He clearly understood that two waves, in order to produce interference, had to satisfy various conditions which today are reduced to two, namely that they must be homogeneous and that they must be coherent, that is to say they must have equal wavelengths and a constant phase difference. In practice this meant that, to obtain interference, it was necessary to produce an overlap of waves emitted by a single source, after they had travelled along different optical paths. When Fresnel began to experiment with double refracting substances he was forced to realize that the overlapping of beams emitted by the same source, but which had been separated into an ordinary and an extraordinary beam by a crystal, never produced interference, irrespective of how much the difference between the optical paths was varied. He concluded that two beams differently polarized could not influence one another, but it was necessary to explain this behaviour. In the famous prize-winning memoir presented to the Académie des Sciences he had on several occasions expressed his views concerning the structure of light waves. They had to be elastic modifications of the ether, that elusive and rarefied fluid which filled the universe and permeated even the pores of material bodies. As it was a fluid, indeed the supreme fluid, according to the mechanics of the period, the vibrations could only be longitudinal, for it was thought that transverse vibrations could only be present in solid bodies. Fresnel built up the whole of his theory of diffraction by considering the vibrations of his waves to be longitudinal. When, however, he was faced with the disappearance of interference in the combination of an ordinary and an extraordinary beam, he was forced to conclude that the waves could not be longitudinal but had to be transverse. In this he found the very mechanical quality which characterized the phenomena of polarization. In fact, longitudinal vibrations showed a symmetry of revolution about the ray, whereas transverse vibrations showed a symmetry with reference to two planes, which passed through the ray, at right angles to each other, namely, the plane of vibration and that plane perpendicular to it. This was indeed the symmetry shown by the phenomena of polarization.

Fresnel did not hesitate to accept this structure, particularly since it explained fully how two beams, one ordinary and the other extraordinary, did not produce interference. Since two vibrations at right angles to each other could not produce interference, it was enough, for the theory to correspond perfectly with the experiment, to imagine that the ordinary and the extraordinary beams were polarized in two planes at right angles to one another.

This concept of transverse waves met with the greatest hostility from the scientists of the day, who could not imagine an extremely fluid and rarefied ether which at the same time possessed the mechanical properties of a rigid body. Even Arago admitted that he could not follow the exuberant engineer in his ideas. But Fresnel was convinced that at last he had the key to many mysteries, and with his model of waves he gave a final clarification of the phenomena of polarization. With insuperable precision he explained a long series of extremely complicated experiments, such as those of chromatic polarization that Arago himself had discovered by chance in 1811, and which the followers of Newton could not explain in spite of all their efforts. Following this line Fresnel reached the synthesis which is his masterpiece. By studying the phenomenon of double refraction in uniaxial and biaxial crystals he concluded that the more general form of a wave front in a material medium is that of a surface of the fourth order, which could degenerate into a sphere and an ellipsoid in uniaxial crystals, and into a double sphere in isotropic media.

To end this brief review of Fresnel's work we must recall the final interpretation that he gave of the famous phenomenon of partial reflection by transparent surfaces, that simple phenomenon which until then had puzzled Grimaldi, Newton, and Huygens, and which in Malus's experiments had unexpectedly acquired a special importance as it had been compared to the great mystery of double refraction. It is our duty to recall that the first proof of the phenomenon of partial reflection was given by Young, who after the early successes of the young French engineer, had found new courage and had started his work once again, collaborating very successfully, thanks to his ability and his imagination, with Arago and Fresnel. In *Lecture XXXIX* of Young we find the following passage:

> It may, therefore, safely be asserted, that in the projectile hypothesis, this separation of the rays of light of the same kind by a partial reflection at every refracting surface, remains wholly unexplained. In the undulatory system, on the contrary, this separation follows as a necessary consequence. It is simplest to consider the ethereal medium which pervades any transparent substance, together with the material atoms of the substance, as constituting together a compound medium denser than the pure ether, but not more elastic; and by comparing the contiguous particles of the rarer and the denser medium with common elastic bodies of different dimensions, we may easily determine not only in what manner, but almost in what degree, this reflection must take place in different circumstances. Thus, if one of two equal bodies strikes the other, it communicates to it its whole motion without any reflection; but a smaller body striking a larger one is reflected, with the more force as the difference of their magnitude is greater, and a larger body, striking

a smaller one, still proceeds with a diminished velocity; the remaining motion constituting, in the case of an undulation falling on a rarer medium, a part of a new series of motions which necessarily returns backwards with the appropriate velocity; and we may observe a circumstance nearly similar to this last in a portion of mercury spread out on a horizontal table; if a wave be excited at any part, it will be reflected from the termination of the mercury almost in the same manner as from a solid obstacle. . . .[11]

Thus the mysterious phenomenon lost all its difficulty with the advent of the wave theory. In Fresnel's hands the explanation became final and complete.

The intervention of the polarization of the incident beam, of the angle of incidence, and of the index of refraction of the two media in contact were defined precisely by what we now call 'Fresnel's formulae', which are today still considered to be valid, having been confirmed by nearly a century and a half of experimental tests.

In 1824 Fresnel's health, which had always been delicate, deteriorated to such an extent that he could only attend to the official duties of his employment and he was compelled to renounce all other work. At the beginning of 1827 he was compelled also to resign from his employment, and on July 14th of the same year, he died. He was then barely thirty-nine years old.

In the first half of the nineteenth century there took shape a mechanical model of light with the following characteristics. In an ether which was very fluid, elusive, imperceptible, and endowed with the mechanical characteristics of a rigid body, pervading the whole universe and all the pores of matter, waves were propagated having a wavelength between 0·8 μ and 0·4 μ, with a velocity of 300,000 Km per second in vacuum, implying a frequency of about 5×10^{15} cycles per second. This vibration was of an elastic nature and completely transverse. When all the vibrations occurred parallel to a plane which in its turn was parallel to the direction of propagation, the wave was said to be polarized and showed a sym-

[11] T. Young *A Course of lectures on Natural Philosophy and the Mechanical Arts* London 1807, Vol. I, Lecture XXXIX, p. 461

metry of behaviour with reference to the above mentioned plane. Beams consisting of many waves, each one of which was polarized with reference to a plane of its own, but different from wave to wave, no longer presented such a symmetry, but rather showed a symmetry of revolution about the direction of propagation. Such beams were called beams of 'natural light'.

The vibrations were present in all matter since they were propagated in the ether which was contained in the matter, and the mechanical characteristics of the ether were modified by the matter itself. As a result, a double phenomenon took place at the surface of separation or of discontinuity, between the zones of pure ether and the zones of the ether mixed with matter. Some of the vibrations travelled back in the pure ether, following the laws of reflection, while others penetrated the zone of ether mixed with matter. The proportions of the two parts were determined by various factors, such as the wavelength, the state of polarization of the incident light, the angle of incidence, and the nature of the material zone. In each case, in this material zone, the velocity of propagation of the vibration was a function of the wavelength, and this involved refraction and dispersion. Similar phenomena took place when the vibration passed from one material medium to another endowed with different characteristics.

The propagation of light in an homogeneous medium was ruled by laws common to all waves, according to which every point struck by the vibration became the centre of secondary wavelets all having the same phase. If the medium was also isotropic, then the waves were spherical and had for centre the point source; if the medium was anisotropic, then in its general form, the wave front was a surface of the fourth order; when the medium had an axis of symmetry, then the wave front degenerated into a sphere and an ellipsoid. When the waves met obstacles or when the wave front was altered by discontinuity zones, the application of the fundamental principles of their propagation led to more complex results which together formed the group of phenomena of diffraction. When two waves that were coherent, homogeneous, and not polarized at right angles to each other, overlapped, interference took place; the combined result was not always the arithmetical sum of the pulses taken separately, but could in fact be smaller, if not zero, according to the relative phase of the two motions. Waves of different wavelength excited in human eyes sensations of different colours. The shorter the wavelength, the more the colour was displaced towards the violet in the sequence of colours of the spectrum.

We must admire such a magnificent synthesis. Yet the supporters of Newton did not surrender nor did they lose hope of reinstating the concepts they had learnt in their youth. They were deeply attached to these concepts which they had discussed and accepted with great confidence, great faith, and great admiration. Indeed these concepts had almost become part of them, and they could not jettison them. The secret hopes of the supporters of the theory of emission were encouraged by the difficulty that mechanics itself created when transverse vibrations were considered and by the problems that the concept of the ether raised at every turn, and which required a continuous revision of the current concepts concerning the constitution of matter and the nature of physical space.

In 1849 Fizeau (1819–1896), and in 1862 Foucault (1819–1868), succeeded in carrying out the first terrestrial measurement of the velocity of light, the former by means of a toothed wheel, and the latter by means of a rotating mirror which allowed the experiment to be performed in a laboratory. By allowing the beams of light to travel through long tubes filled with liquid it became possible to obtain direct proof that the velocity of light was smaller in liquids than in air or in a vacuum. Confronted with this experimental proof, even the most ardent follower of Newton had to admit defeat.

CHAPTER EIGHT

What is Light?

S

It was an undeniable fact that the wave theory predicted phenomena of diffraction, of interference, and of polarization as exactly as the theory of universal gravitation predicts the motion of celestial bodies. Indeed the fact that this theory predicted the three groups of phenomena which since their discovery had appeared as surrounded by unsurmountable difficulties, produced in physicists the certainty that the wave theory was nothing less than the 'truth'. All this strengthened the certainty, already deeply felt among physicists, that physical reality was an entity which could be grasped and understood. It is true that some physicists had reservations on this subject, but they were considered to be abstruse philosophers, remote from reality and experiment. In schools, where the mind of the young, hence general opinion, is formed, there was taught the indisputable idea that 'light' was a collection of waves with wavelengths ranging between 0.4 μ and 0.8 μ, and that the individual waves, from the shortest to the longest, constituted the sequence of colours of the spectrum from violet to red.

Substantially this assertion was very similar to that of Newton, the only difference being that waves of given wavelengths now took the place of corpuscles of given dimensions. The confusion started by Newton himself when he said that he called 'red making' the rays which enabled us to see a red colour, was so advanced that by now no one went back to the origin of the idea; general opinion had become accustomed to the idea that the red rays of Newton were almost like a fine red powder, propagating at a speed of 300,000 Km per second, and that red was always this powder. When waves took the place of corpuscles, this materialistic concept of colour became grotesque, but the habit of considering red as something physical, external to the observer, had become so ingrained and natural, that people continued to speak of red, yellow, green, blue, and violet waves. Moreover, now the wavelength could easily be measured, it was an experimental fact, and, therefore, colour appeared as a physical characteristic of light, while in the corpuscular theory, colour was purely a hypothesis because no one had measured the mass of the various corpuscles. Thus the possibility was eliminated of considering light and colour to be anything else. The young minds of students in schools were trained to accept this first and simple idea that light and colour were physical phenomena, the mechanism of which was completely understood, so that the unfolding of the more complicated experiments could be predicted in their minutest detail. The study of light and colour was pursued indefatigably in order to define even better the nature of these entities and their laws.

We shall follow the development of these investigations in three directions. The first concerns the studies on the nature of light, the second concerns the measurement of light, and finally the third concerns the measurement of colour.

In the first half of the nineteenth century, Fresnel's waves were elastic and transverse, and the medium in which they were propagated was the ether, that mysterious fluid which pervaded the universe and permeated material bodies. To explain a considerable group of phenomena of light, the wave theory required that the vibrations should be transverse, and this led to attributing to the ether very strange properties as we have already described. Its relation to matter, the possible rest or movement either total or partial with reference to it, became at once the subject of scientific investigation which did not give worth-while results.

A great change, however, took place in the second half of the nineteenth century, when the development of the studies of electromagnetism reached a peak with the challenging theory of Maxwell (1831–1879). In 1873 Maxwell advanced his theory that light waves were electromagnetic waves and, apart from wavelength, they were identical with all the waves which could be obtained by radiation from electrical circuits which had suitable inductance and capacity. The experiments of Hertz (1857–1894), carried out in 1888, marked the beginning of the 'optics of electromagnetic waves', so called because these waves obtained by means of oscillating electrical circuits, appeared to show effects of reflection, refraction, diffraction, interference, and of polarization and to obey the same laws which in optics had been formulated for light. The result of this was that light waves were no longer considered as elastic waves but as electromagnetic waves. However, this simple solution was not satisfactory for very long. The new ideas concerning energy led physicists to consider light waves from this point of view, and the interaction between light and matter both in emission from luminous sources, and absorption by illuminated bodies, became the object of countless theoretical investigations and measurements, and the conclusion was reached that the general picture was not satisfactory.

In 1900, thanks to the work of Max Planck, the concept of quantum of energy was introduced. In 1905 Einstein (1879–1955) extended this concept of discontinuity to luminous energy, and showed how, by means of 'quanta of light', which N. G. Lewis in 1926 called 'photons', it was possible to explain a large group of phenomena which could not be reconciled with the classic wave model. Thus the great battle between waves and photons began. It so perplexed physicists that they were led to reconsider the fundamental concepts of science, and even to re-examine thoroughly its possibilities and its purposes. This dissension took place among physicists at the same time as the theory of relativity was questioning the concepts of space and time; it destroyed the euphoria which in the nineteenth century aroused such confidence in scientific research and in experimental investigations.

We must leave this stormy period without discussing it. To enter into the details of such an imposing event would take us too far from our own field.

The measurement of light continued very slowly along the path indicated by Bouguer, whose work we have discussed briefly in Chapter VI. The faith and tenacity shown by those who dedicated themselves to this work are to be admired. Their behaviour is also noteworthy because it stands as a striking example of the strength of prejudice, even in experimental investigations which claimed to be objective and impersonal as far as possible.

It is obvious that the idea of measuring light could only arise in the mind of a person who was convinced that light was a measurable entity. To be measurable this entity had to be objective and external to the observer, because, had light been considered as a subjective and psychic entity, the observer would not have considered measuring it by physical means.

In physics there is another field which proves what we have just said. In the case of heat, in fact, a clear distinction has been made between 'heat', and 'hot' or 'cold'. Physicists have agreed that hot and cold are subjective entities, they are sensations felt by an observer. Some feel one way, others in a different way, thus one may feel 'hot' while another feels 'cold', indeed even under the same external agent the same observer may sometimes feel 'hot' and other times 'cold'. This, therefore, is of no interest to physics, which is only concerned with what exists by itself independently of the observer. The external agent which makes a living body feel 'hot' or 'cold' was called 'heat', and only this was considered by physicists. They soon discovered that man, as a detector of heat, was a very unsatisfactory instrument, and, moreover, he had a very limited field, therefore they introduced thermometers, bolometers, and thermocouples, and in physics there is no place for 'hot' and 'cold' any more. This, in effect, is the correct reasoning of physicists. It is strange, however, that this clear and unequivocal method was never applied to light or to sound. This was probably due to the fact that light was not thought as being something of the same nature as 'hot' but rather that it was something of the nature of 'heat'. It is thus that photometry was born and developed. Photometry advanced hesitantly because it was based on a very arguable concept. It is only in the last few decades that photometry has followed a reasoned path after many years as a temporary and very indefinite technique.

As long as it was a question of carrying out relative measurements, that is measurements which expressed the percentage of reflected light, of transmitted light, and of diffuse light, there were no difficulties because it was a question of physical measurements. Even if the eye was used as an instrument for detection and comparison, there were no complications, because the eye was only asked to give a judgement of equality. From the times of Bouguer, as we have already said, the methods of photometric measurements were purely 'null methods', because it was realized that no observer could give a reliable judgement concerning the ratio of two luminous intensities. This meant that when an observer 'saw' a given intensity, he acted under the influence of personal reactions which could not be understood or controlled and which, under the stimulus of the same external agent, sometimes gave one answer and sometimes another. These anomalies and uncertainties were attributed to the functioning of the eye, and physicists left this field to be studied by physiologists and psychologists. Meanwhile, to eliminate the difficulty, they decided to refer to a 'normal eye' or 'standard observer', namely to a fictitious eye which satisfied certain conventions and which was invariable. Naturally numerous measurements were made in order to choose this 'eye', so that it should be as near as possible to the greatest number of real eyes, hence near to the average of them. By this method photometry became independent of the vagaries of the physiology and of the psychology of vision, and continued on this path with the conviction that the measurements were physical measurements of physical light. For a long time it became a question of white light photometry, monochromatic light photometry, and heterochromatic light photometry. The only real one was white light photometry, but unfortunately no definition existed of 'white light'. When an attempt was made to find such a definition, physicists were compelled to reach the conclusion that 'white light' had no definite physical significance. In addition there appeared to exist a very great number of white lights. Light which seemed white in certain conditions could, under other conditions, assume a definite colour, and it was still the observer, affected by physiological and psychological factors, now made more complex by the question of colour, who had to make the decision whether the light was white or not. As a result of all this it became expedient to define white light following certain conventions, indeed more than one definition was proposed. The definition was made by means of standard sources called A, B, and C, with the assumption that white light photometry was to be carried out with ordinary sources, even if these were not typical sources, and accepting the results which could be obtained from the measurements and which were in some cases uncertain. Experiments showed that the various white lights could be compared with one another and could hence be measured even when differences in the hue existed, with the understanding that the accuracy of the measurements was greater when the differences were smaller. In all this we recognize a compromise, and we admire the tenacity of the physicists who made every possible effort to grasp something which seemed continuously to escape them. For a long time monochromatic photometry was considered to be of restricted interest, and heterochromatic photometry so difficult that it had no practical use. We shall not stop to trace further developments because this branch of photometry assumed considerable importance when it became known as colorimetry. We shall return to this subject shortly.

In ordinary photometry, that is white light photometry, the standard of intensity was defined, as well as the photometric units, namely the quantity of light, flux, intensity, and luminance. Leaving aside illumination, we would like to call the attention of the reader to the meaning of these quantities. In text books, 'quantity of light' is defined as being equal to the luminous flux multiplied by the time during which the emission takes place. Flux is the rate of flow of light energy from a source. Intensity is the number of candles of a source determined by comparison with the fundamental standard. Therefore, to obtain the 'quantity of light' emitted by a source, we must determine, by means of a photometer, its intensity by comparing it with a standard source. The number of candles thus obtained is multiplied by the given solid angle within which the emission takes place, and the flux, expressed in a given number of 'lumen', is obtained. The flux is multiplied by the number of hours during which the source remains lit, and a product is obtained called 'lumen-hour', which measures the quantity of light emitted by that source, within that given solid angle during all that time. This is a light emitted by the source, a light which has travelled within a given solid angle and, therefore, is a light which has existed irrespective of whether any observer received it or saw it. It is an 'external light', and this is the light as understood in photometry, But if externally to us there exist waves, or corpuscles, or photons, are these 'light'? This is the crucial point. The source emits waves or photons, but these are not 'light'. They are movement, energy, or even matter if we wish to include the corpuscular theory, they are 'something red making', but not red, because in order to become red that 'something' must reach the retina, stimulate it, and make it transmit impulses to the brain, to the psyche, and it is there that 'red' will exist but not before. What has been said for red is true for green, for violet, and also for white. The subtle transition from 'red making' to 'red', introduced by Newton himself, caused the ambiguity which has lasted for centuries and that still undermines the foundations of photometry. Naturally even the most deep-rooted ambiguity, when confronted with scientific criticism, ends by being isolated and clarified. Today photometry is consolidating its own foundations with rational elements. But before coming to the final conclusions, we ought to review briefly the evolution of colorimetry.

Without examining again the long and difficult work of the ancient philosophers about the concept of colour already discussed in previous chapters, we shall start from Grimaldi's assertion that colours were not 'as generally believed' something distinct from light and existing on the bodies. Grimaldi attributed such importance to his statement that he introduced it in the title of his Book II. Indeed he advanced the idea that colour was due to a vibration of the fluid constituting light, thereby anticipating, even if in a rudimentary form, the concept that later was to be found in the wave theory. One of the greatest merits of Newton was his discovery of the relation between the refrangibility of the various rays and their colour. The discovery of the dispersion produced by prisms and the wealth of experiments carried out with them to separate complex light into elementary indivisible components, followed by the synthesis which by recombining the elementary components enabled the original light to be re-obtained, led to the belief that the problem had been wholly solved and that nothing concerning colour could be added after Newton.

Newton's followers, by exaggerating in this field as in many others, forgot the reservations which Newton himself had made, even if only briefly expressed. In Proposition VII, Book I, Part II of *Opticks* he wrote:

. . . I speak here of colours so far as they arise from light. For they appear

> sometimes by other causes, as when by the power of phantasy we see colours in a dream, or a madman sees things before him which are not there; or when we see fire by striking the eye, or see colours like the eye of a peacock's feather by pressing our eyes in either corner whilst we look the other way. Where these and such like causes interpose not, the colour always answers to the sort or sorts of the rays whereof the light consists. . . .[1]

Naturally the beauty and the importance of all the studies made on the subject of colours 'which have their origin in light' attracted great attention, and it is not surprising that no interest was taken in the other colours originating in abnormal and rare circumstances. The fact that these also existed did, however, attract the attention, though much later, of men who did not belong to the academic world. We have to reach the beginning of the nineteenth century before we can find someone who had the courage to criticise Newton.

We ought to draw attention here to the strange action of Goethe (1749–1832) who, in 1810, published a large volume consisting of 1,400 pages, entitled *Zur Farbenlehre*, in order to attack Newton violently and to defend the subjective nature of colour against the irrational and overbearing attitude of Newton's disciples. These had reduced the whole subject to a simple physical mechanism, and did not tolerate doubts or reservations of any kind on the subject. They observed an absolute silence on everything which could in any way inconvenience their way of thinking. Goethe must have been strongly convinced, for he fought this battle as if he were the only sighted man in a world of blind people; only in this way can we explain and justify his action. He could not have hoped to gain any advantage by devoting so much time and energy in writing such a lengthy book. He was, it is true, a famous writer in the literary field, but he certainly was at a disadvantage in a struggle with Newton in a scientific field. The fact that he did it, showed that he was convinced that he alone could undertake such a task, and, therefore, he felt that it was his duty. No one shared his views on the subjectivity of colour until his voice was heard and understood by Schopenhauer (1788–1860), another famous philosopher, who in

[1] Newton Op. Cit. Book I, Part II, Prop. VII, p. 443

1816 in a small volume entitled *Ueber das Sehen un die Farben* strengthened Goethe's concepts and added some improvements. We shall not discuss in detail the criticisms and the views of these two philosophers because their attempts had no sequel. In that period the followers of Newton had other worries. In their midst there was evidence of an imminent and serious danger that before long was to destroy them, and, therefore, they were more concerned about this than about a challenge from seemingly innocuous strangers. As for their successors, they were only beginners, and they were so busy in asserting the fundamental principles of the wave theory that they were not interested in any other complex questions. Goethe, in his book, took Newton to task and literally quoted the text of several passages of *Opticks*, in order to criticise them step by step, and therefore the followers of the new theories were not involved. This was an unfortunate situation in which the voices of the two philosophers failed to have the effect they would have had at some other time. Even if the time had been more favourable, the effect of Goethe's protest would have been only modest, because his was a call for a return to the past, to subjectivism, at a time when objectivism ruled with exaggerated power. This was the time when the 'new natural philosophy' was celebrating the annihilation of classical philosophy, and no one was interested, not even in those ideas that were good and indestructible. The return to the past in the eighteenth and nineteenth centuries was looked upon with pity as being foolish. For this reason even when Young studied the question of colours he did not follow a line similar to that expressed by Goethe.

Young, as we know, had studied medicine, and he was endowed with a rebellious and anti-dogmatic spirit, thus he was not a physicist and he was not tied to the Newtonian school. On the contrary, once started on the wave theory as far as the nature of light was concerned, he became decisively anti-Newtonian; and then he began to study in great detail the theory of colours. Naturally, because of his training and his background, he approached the whole problem more as a physiologist than as a physicist, but with a tendency more to physics than to psychology. In other words he centred his attention on the important part played by the eye, considered as the organ of the sense of sight, in the formation of colours; and he decided to study and to define this role without delving too far into the more obscure question of the psyche of the observer. This direction of studies was very fruitful because it did not meet with the hidden but general hostility of the scientific world of the nineteenth century. It developed, though slowly, into what is known in modern times as colorimetry. We owe to Young the fundamental concept of the 'three colour system'. He noted that colours were not determined simply by the physical composition of the beam of corpuscles or of waves striking the eye, but that the same colour could be seen by sending to the eye beams of very different compositions. Indeed, by taking three fundamental colours and by mixing them in suitable proportions, it was possible to obtain an unlimited range of hues and tones, so that it could be said that all possible colours could be obtained, including certain colours that, like the purples and the magentas, were not to be found among the colours of the spectrum. This way of thinking shifted the centre of the question from the structure of the beam of light to the structure of the retina. In other words the physiological approach prevailed over the physical.

Young advanced the hypothesis that in the retina there were three types of receptors, each one corresponding to one of the three fundamental colours, and that the sensation of one colour rather than another was the result of how the three groups of receptors were stimulated. If two beams of different physical composition acted equally on these receptors, the result was a sensation of the same colour in either case. Faced with considerations of this nature, physicists thought it natural to ignore them and continued to consider 'their' colours, leaving to the physiologists the task of investigating the functioning of the eye. On the other hand the philosophical influence of the Newtonian school had been so deep that the 'objective' direction of the studies continued to rule without discussion, even when waves took the place of material corpuscles. The three colours theory, however, did not disappear. It was taken up again by Grassmann (1809–1877) and by Maxwell, and became of great value in the hands of another famous scientist, H. L. Helmholtz (1821–1894), who, having studied medicine and physiology and having a very wide and varied scientific culture, was best qualified to enter the field of studies relating to colour. The trichromatic theory had a great development thanks to his work and that of his gifted collaborator Arthur König. As happens to all theories when they are strained beyond their capabilities, Young-Helmholtz's theory, carried to extremes, began to show its weaknesses, especially when faced with experimental evidence.

The number of students of colorimetry increased rapidly as the nineteenth century was ending. The experimental data collected were considerable, thus conclusions and conventions were reached which were approved with international agreement by the International Commission on Illumination. The lines followed could be summarized thus: the trichromatic theory was accepted as a fundamental basis, even though it was acknowledged as imperfect. It had to be admitted that one particular group of three fundamental colours did not exist, but that the choice of the three colours which, by varying their proportions could be used to obtain all other colours, was arbitrary, practical considerations apart, within wide limits. This led to the result that Young's hypothesis concerning the triple structure of the sensorial system of the eye did not correspond to reality. In the question of colour, however, the number three has kept a pre-eminent character. The conclusion was reached that, to define a colour, three parameters were necessary; these could be the percentage of the three fundamental colours, but could also take a different form. One of the most interesting conclusions reached by the International Commission on Illumination was that every colour is characterized by its brightness or luminance, by its hue and by its saturation. Brightness is the attribute of a body that enables us to distinguish when a given colour is lighter or darker, according to the degree of illumination. Hue is the attribute that we usually express with terms such as red, green, yellow, and so on; and saturation is the attribute which enables us to differentiate between red and pink or between bottle green and pale green. Every colour is considered as a mixture of a pure colour either 'spectral' or 'saturated' and of a certain amount of 'white' which attenuates the saturation. Given any mixed colour, the 'dominant wavelength' has been defined as that of the corresponding spectral colour. Thus it became possible to suggest methods which could be followed, and instruments which could be constructed, to measure colour. With delicate manipulation, very simple in concept, it became possible to determine the dominant wavelength, the saturation and the brightness of a given sample, once the spectral composition of the source of light which illuminated it had been established. The students of colorimetry wished to measure the colour of a body independently of the observer, in the same way as quantities are measured in physics. They concluded that the chromatic effect, namely what an observer saw when he looked at a given body, depended on three factors:

1. The spectral composition of the illuminating light.
2. The physical properties of the body under examination, reflection or transmission.
3. The sensitive properties of the eye of the observer.

In essence it was a question of determining the second factor, which was the one determining the colour and, therefore, it became necessary to stabilize the other two factors. The colour of a body appeared different when the illuminating light was changed and when the observing eye was changed. Indeed sometimes the colour appeared different even to the same eye on account of numerous physio-psychological causes, such as fatigue, adaptation of the eye, imagination, or fear. If it was a question of finding three factors which could define the colour of the body, it was necessary to establish with which source of light and with which eye the measurement was to be made. This led to a definition of the illuminating source, as we have mentioned, and to a definition of the eye. An international convention established the curve of chromatic sensitivity of the eye to pure spectral colours, as a function of the wavelength of light. However, in visual determinations there was the intervention of the eye which could be, and indeed in general was, different from the conventional eye, so that 'impersonal' methods, that is to say completely objective methods, were adopted. By means of these it became possible to determine the three characteristic factors of the colour that the conventional eye would see of a body illuminated by the conventional illuminating source. All this is very neat and simple, but it clearly means that we must give up any attempt to measure colour. It was inevitable that this conclusion should have been reached, for colour, like light, is not a physical entity and therefore cannot be measured by objective means and methods.

At the beginning of the nineteenth century, during which science made so many new conquests, even before Young and Fresnel had achieved their great revolution, other observers had detected strange phenomena. A thermometer exposed to a beam of sunlight registered an increase in temperature, indicating the existence of heat. This was not a new experiment, in fact several times in previous chapters we have mentioned the discussions that took place concerning the fact that bodies exposed to light became heated; indeed it had even been remarked that the heating was different according to the colour of the body which was exposed to the light. At the beginning of the nineteenth century it was desired to express this phenomenon in more precise terms, and measurements began by using a thermometer instead of a body.

William Herschel (1738–1822) went further. He placed several thermometers close together on a plane, and projected on them the solar spectrum, in such a way that each thermometer received only a small portion of the spectrum. Some thermometers happened to be beyond the coloured and bright band of the spectrum, and a new effect was detected. The thermometers exposed to the radiation of the solar spectrum registered temperature. Those exposed in the red part showed a higher reading than those exposed in the green or violet part of the spectrum, but the unexpected finding was that the thermometers which were exposed outside the visible spectrum, on the side of the red or beyond the red, registered an even higher reading. Therefore there existed rays which were not visible, but that a prism could refract in a similar way to light rays. It was deduced from this that in the beams of sunlight, in addition to red, yellow, green and violet light which when mixed produced white light, there was also an 'invisible light'. Just like visible light, this invisible light had its own refrangibility; since this newly discovered light was less deviated by the prism than the red, its refrangibility was less than that of all visible radiations. Thus the infra-red was discovered. On the other hand, centuries earlier, alchemists had noticed that certain silver salts, such as silver nitrate and silver chloride, became black when exposed to sunlight. In 1777 C. W. Scheele (1742–1786) had shown that different monochromatic lights did not have the same effect on these salts, and that the most active was the violet light. In 1801 Ritter (1776–1810) and then Wollaston (1766–1828) noticed that if the solar spectrum was projected on a plate of silver nitrate, the darkening effect took place even where the eye could not distinguish any colour, but only on the side of the violet. Beyond the violet, therefore, there existed a light more refrangible which manifested its presence by means of its action on silver salts; although it was not visible to the human eye, this too then was an 'invisible light'. Thus the existence of the ultra-violet was discovered.

With the advent of the wave theory, the 'invisible lights' could be explained. Following the experiments on interference of Young and of Fresnel, it was possible to measure easily the wavelength of coloured light. The conclusion was reached that red radiation had a wavelength of about 0.8 μ or 0.7 μ, yellow radiation a wavelength of about 0.6 μ, green radiation had a shorter wavelength, and the violet radiation had a wavelength of 0.4 μ. It was simple enough at this stage to suggest the existence of radiation of a wavelength longer than 0.8 μ for the infra-red and of a wavelength shorter than 0.4 μ for the ultra-violet, and because to the values between 0.8 μ and 0.4 μ there corresponded the various colours of the spectrum, the new waves, both the longer and the shorter, were called 'black lights'. These new discoveries did not seem to attract much attention. No one seemed to realize the revolution that these discoveries were introducing in the concepts of light existing at the time. No one noticed that to speak of 'invisible light' and of 'black light' was a contradiction in terms. Nevertheless investigations in these new fields continued, and indeed increased rapidly, helped considerably by the introduction of the thermocouple as a detector of the infra-red, and of the photographic process as a detector of the ultra-violet. These investigations had the effect of extending even further the limits of the wavelengths which could be detected. Light, however, was still considered as an entity consisting of a central group of waves, both visible and coloured, and of two groups, one at either side, consisting of waves that were both invisible and black. Maxwell's intuition followed by the experiments of Hertz, unified these three groups into one which extended to waves having a wavelength measurable in kilometres. The elastic nature of waves was finally replaced by the electromagnetic one, and this new field was extending in a manner which could not have been imagined. In 1896, eight years after the early experiments of Hertz, Roentgen discovered X-rays, and this was soon followed by the discovery of γ-rays of radioactive substances. After research lasting almost twenty years, the experiment suggested by Laue and carried out by Knipping in 1914 finally led to the conclusion that these rays were new groups of electromagnetic waves with extremely short wavelengths of the order of one thousand millionth of a millimetre or perhaps even less. This synthesis, which practically covered the whole range of electromagnetic waves from a wavelength of a few kilometres to a few thousand millionth of a millimetre, was really imposing. It represented one of the greatest achievements of natural philosophy, even if later many doubts were raised and reservations were made about the wave structure of the whole complex system. We shall not dwell on this subject because what interests us is the effect that all this had on the question of the nature of light.

Formerly light was a whole in the sense that all the group of known rays was made up of visible entities, luminous even if coloured, and by them only. This led scholars to accept an implicit hypothesis not totally unjustifiable, at least as a temporary measure, that between the light seen *lux* and the external agent *lumen* there was a correspondence which, at a first approximation, could be considered as an equality or at least a proportionality. Rightly or wrongly *lux* and *lumen* were, therefore, united into a single thing, 'light'. After all, this was only a working hypothesis which in the field of science had as much right as any other hypothesis. It would have been better to have stated it explicitly, but since this was not done we must accept the events as they happened. When, however, the existence of 'black light' had to be admitted, complications ensued, because it became obvious that here there was a *lumen* to which no *lux* corresponded, and the above mentioned hypothesis did not hold. By now the custom of considering only *lumen* as an entity of interest had become so general, at least among physicists, that they forgot that it was only a temporary hypothesis which enabled *lumen* and *lux* to be considered as one, and they continued to call the whole 'light'. When, however, the 'dark light' was extended to radio waves, and included X-rays and γ-rays the situation became too complicated. The solution adopted was very simple: the situation was ignored, the definition of what was meant by 'light' was avoided, and the old language was still used. When anyone asked for a clarification and called attention to some of the numerous contradictions which resulted from upholding these fundamental ambiguities, the answer was: 'It is only a question of words and it is not important'. On the contrary, it was important. It was not a question of words, but rather a question of facts and of basic concepts which had a deep and remote origin in the long history whose most important and obvious stages we have described in this book, and which obviously have exercised the minds of the greatest thinkers throughout the ages. This is why at the end of this history which we have recalled, we ask the question 'What is light?' We feel that a final general picture of the question as we see it today will be useful to the reader who, having followed the evolution of ideas during the last twenty-four centuries, is in a particularly favourable position to appreciate its value.

Man began to take an interest in light as a consequence of the fact that he could see. Thus the notion of darkness and of light was formed as being two conditions of which one was the absence or the negation of the other.

For a long time man attempted to understand how he saw. From the beginning it became apparent that it was the soul, the mind or the psyche of the individual, that did the seeing, and the problem consisted of defining how the psyche could perceive the form, the colours, and the position of the objects seen even when they were far away. First it was thought that rays were emitted from the eye, and went forth to explore the external world. Then it was thought that the *eidola*, which came from the object, entered the eye. No mention was made of light as an entity in its own right. The question of finding out what light was, was not asked, although in the early ideas of Lucretius there can be found the beginning of the assertion that a physical 'something' existed which intervened as a determining element of vision. This, however, was only an indefinite and rudimentary idea.

The Arab School, particularly Alkindi and Alhazen, had established that the eye functioned by the action of an external agent, which carried the form and the colours of the objects observed. For several centuries after, attempts were made to reconcile this concept with the *eidola*, thereby producing complex and unsatisfactory models. In these, however, there began to appear the concept of something rectilinear, of a luminous ray, necessary to guide the *eidola* to the eye, and at the same time capable of reducing it to the necessary dimensions so that it could enter the pupil. This can be said to be the first interpretation of *lumen*; but generally it was added that once the *eidola* entered the eye it was felt by the retina and something flowed along the optic nerve to the brain and as far as the psyche, whose task was to create the figures which were seen as phantoms in their place with the form and the colours carried by the corresponding *eidola*. These figures were the *lux* that was seen and created by the psyche. It had also been discovered that the psyche could see without an external action, because it could create phantoms of its own, as in dreams, in hallucinations and in illusions.

Kepler found the key to the mechanism of vision, and so the *eidola* disappeared, but the general idea remained. A geometrical communication was established between the elements of the observed object and those of the retina, a communication which consisted of cones of rectilinear rays each one having its apex in a point of the object, and which were transformed by the optical system of the eye into other cones each one converging into a point of the retina. These rays, provisionally only rectilinear and geometrical, constituted the *lumen*, and were the agent which carried to the retina the forms and the colour of the external bodies. From the retina, something had to flow along the optic nerve as far as the psyche because, from the analysis of the stimulus on the retina and from the efforts of adaptation and convergence made by the eye in order to see sharply and to avoid double vision, the psyche had to deduce the location of the source of the rays, that is the locus of the apexes of the cones which had reached the pupil. It was there that the psyche had finally to locate the figures, the phantoms which it had created. The explanation of the figures seen behind mirrors, beyond prisms and lenses was the most admirable and the most convincing of the proofs that the mechanism thus evolved corresponded to reality.

Lux was still confined to the psyche, it was still a psychic phenomenon; but *lumen* was now a physical phenomenon partly defined and partly to be defined. Both were linked by a physiological link, because the eye was necessary in order to pass from one to the other. Vision thus assumed the character of a multiple process in which took part an external physical factor which was *lumen*, a physiological organ which was the eye, and an entity that was mysterious but necessary, the psyche. To the latter was also left the task of creating *lux*. The whole problem of defining the nature of *lumen* was now set. The prevailing opinion was that *lumen* consisted of material corpuscles, and this hypothesis was fully developed and acquired much credit. After Grimaldi had shown that colours were a modification of *lumen*, Newton advanced his new idea, namely that colours were not really a modification of *lumen* but were only the physio-psychological effect of the action of the different masses of the corpuscles upon the eye of the observer. The beauty of the experiments and of the theories connected with these ideas, in particular the importance that 'natural philosophy' acquired in the seventeenth and eighteenth centuries, shifted all the attention of the scientific world towards *lumen*. *Lumen* became 'light', while every trace was lost of *lux*.

The theory which considered light as consisting of material corpuscles gave place to the wave theory. Soon it was discovered that there were not only waves capable of being detected by the eye, but that there were many others of the same nature which produced no effect on the eye, ranging from the radio waves to the infra-red, from the ultra-violet to the X and γ-rays. It was recognized that it was a question of a form of transmission of energy which was one of the fundamental mechanisms of the universe. This was indeed a wonderful synthesis from the point of view of physics, but was of grave consequence for light. If that group of waves capable of being detected by the eye was called 'light', was it correct to call 'black light' those waves which had no effect on the eye? At the same time photometry, which had come into being for the purpose of 'measuring light', ended by adopting a convention for 'light'. This prompted the question whether it was correct to call 'light' that particular entity which could be determined by experimental and numerical processes dictated by photometric rules. A similar question arose from the conclusions reached by colorimetry, namely whether it was correct to call 'colour' that entity which could be determined by means of experimental and numerical processes following the rules of colorimetry.

To answer these questions it is necessary to describe a general outline of the process of vision as understood from the ideas developed in the many centuries of research. We must imagine ourselves in an external world where there are material bodies endowed with movement and with energy. We must think of this world as being dark, without light and without colour. The various bodies must be considered as clouds of atoms which radiate energy in the form of waves, or photons, of every wavelength. These waves are to be considered as energy which is transferred from one body to another, and, therefore, they are neither luminous nor coloured, they are entirely dark. Those waves which have a wavelength between 0.4 μ and 0.8 μ, although lacking luminosity and colour, are apt to produce certain reactions on the retina when they finally reach it. In any case we would not be able to grasp the meaning of 'wave luminous in itself' or 'wave coloured in itself'. The reactions on the retina produce the transmission of nerve impulses from the eye to the brain. In the brain, which is the central sensorial organ, a deep and detailed analysis of these impulses takes place. The intensity, the origin and the complexity of these impulses are considered, and as a result luminous and coloured phantoms are created. More precisely, the nerve impulses produced by one single element of the retina are defined by three parameters which are represented by what we call brightness, hue and saturation. The number of simultaneous transmissions to the central organ is as great as the number of the elements of the retina which are stimulated, and we must remember that in the optic nerve, 500,000 fibres have been counted capable of transmitting impulses independently. Thus the psyche, on the basis of all these triple pulses, builds as many elements which together form the phantom endowed with brightness, hue and saturation of each element. From the fusion of the phantoms obtained by means of the two eyes, and from the intervention of other physiological factors, and particularly psychological factors such as memory and imagination, the psyche succeeds in measuring the distance of the source of waves from the eye. As a conclusion of this long, detailed and wonderful labour, the phantom is placed where the position of the source had been located. Then the *ego* which has created these phantoms, and has placed them around itself, 'sees' the space around it populated with these luminous and coloured figures.[2]

This then is the process of vision in its whole indivisible complexity, consisting of a physical, of a physiological, and of a psychological phase. Light and colour exist only in the psychological phase. They are entities exclusively and absolutely subjective, and therefore they are not part of the external world and cannot be included in the domain of physics. From this conclusion we appreciate how very unfortunate had been the loss of the precious distinction that our mediaeval forebears used to make between *lumen* and *lux*. If we wished to rectify our concepts we ought to re-establish the two terms. The word 'light' ought to refer to *lux*, as the opposite of darkness, because this was the meaning attributed to it by everybody. It could also be suggested that it would be better to establish a different convention, and to reserve the meaning of

[2] For further information concerning the mechanism of vision see V. Ronchi *Optics, the science of vision* New York Un. Press, 1957, translated by E. Rosen;
V. Ronchi *Critica dei fondamenti dell'acustica e dell'ottica* Centro didattico Nazionale per l'Istruzione Tecnica e Professionale Roma 1963.

lumen for the word 'light', as physicists are in the habit of doing. This, however, would require the introduction of another word to indicate *lux*; but since 'what is seen' is called 'light' by everybody, it would be extremely difficult to introduce a new word into common usage. On the other hand if we wished to use a word other than 'light' to indicate *lumen*, it would be easy to use 'radiation', or 'radiant energy'. We reach, therefore, the following conclusion: the physical world is permeated with radiation which has no light or colour; the physical world is black and dark. When those particular radiations which have wavelengths between 0.4 μ and 0.8 μ reach the psyche through the organ of sight, they bring about the formation of luminous and coloured phantoms, namely light and colour.

Let us now consider the 'light' of photometry; is it radiation or light? It is neither: it is not light because we have already said that this is an external and objective entity independent of the observer. Neither is it simply radiation, because it takes into account the sensitivity of the eye, even if a 'conventional eye'. It is in fact a certain operation, a function of radiation, which tries to represent the effect of the latter on a mechanical eye which physiologically does not vary, and which is independent of the psychological mechanism. It could be said to be radiation in so far as it is felt by the 'conventional eye'. Perhaps the simplest way of representing its meaning is to use the expression 'optical radiation'. The expression 'visible radiation' has also been suggested but this is not preferable to the former because no one has ever succeeded in seeing any radiation, and therefore even those radiations which are capable of stimulating the human eye cannot be described accurately as being visible.

Radiation can be revealed by means of detectors or receptors, ranging from radio receivers, to thermocouples, to bolometers, and to photoelectric devices, as well as photographic emulsions and ionization chambers. The eye is also a receptor. Some receptors absorb all the radiation they receive and then they transform it; these are known as 'integral receptors'. Others are selective: they absorb only certain radiations and even the absorption of these radiations may vary according to their wavelength. Among the selective receptors we can mention photocells, photographic emulsions, and also the eye.

The combined intensity of radiation and the sensitivity of the receptor can be calculated by a suitable integral. In the case of a photocell, the integral gives us the intensity of the current developed by the photocell when it is exposed to a beam of a given radiation. In the case of a photographic emulsion there follows a darkening of the emulsion which can be observed after development. In the case of the eye we have an effect similar to the others mentioned above, which could be said to be the alteration of the layer of the retina struck by radiation. In photometry a correspondence between this alteration and the light which the subject would see has been postulated, and the result of measurements and calculations has been called 'light'. We must admit that, in spite of the numerous reservations that we must make on this subject, these various operations are very useful in practice, though naturally difficulties are encountered. It became necessary in this field to distinguish between that which belongs to the realm of physics and that which belongs to the realm of psychology. Recently a new group of parameters has been introduced similar to those used in photometry. This was done to indicate separately 'that which is', and which can be detected by inanimate instruments, from 'what is seen' by a living observer. Thus to indicate the quantity used to represent precisely the 'brightness' seen of a luminous or of an illuminated surface, two terms were coined after long discussions, 'luminance' and 'brightness'. The former indicates a physical quantity as we described it above, and the latter a psychological one. Let us consider the well-known phenomenon of the adaptation of the eye. When the eye passes from a light environment to a dark one, it becomes with time more sensitive, and sees faint lights better. Faint lights, which nevertheless, are due to physical sources of constant intensity. This nowadays can be expressed adequately by saying that brightness increases with time, while luminance remains constant.

Ideas have now been put in order, but light has become a very elusive 'something', so much so that were an enquirer to insist on the question 'What is light?', we would be compelled to confess that nothing definite by itself exists to which this name could be given. If this name cannot be given to the external agent which must be called 'radiation', and if we do not accept that true light is the operation performed in photometry for experimental and technical purposes, then we must search in the realm of the psyche for the 'something' which can be called 'light'. Now, in the realm of the psyche in which we are concerned, we find only phantoms endowed with a given 'brightness' (in the modern sense as defined above), hue and saturation. Nothing exists which flows between psyche and phantom to which the name of 'light' could be given. To the word 'light', therefore, only one meaning remains, namely 'absence of darkness' the very same meaning attributed to it by philosophers two thousand years ago. That 'there is light' simply means that the psyche is not idle but produces phantoms, even if only in a dream.

Index

1450 1500 1550 1600

Other Countries

Snell

GERMANY

Kepler

Marci De Kronla

ENGLAND

FRANCE

Gassendi

Ferm

ITALY

Ronchi, Vasco, 1897-
The nature of light: an historical survey; [revised &] translated [from the Italian] by V. Barocas. Cambridge, Harvard University Press, 1970.
xii, 288p., 53 plates. illus. (some col.), facsims., ports. 23cm. index. (Heinemann books on the history of science)
Illus. on lining papers.
"Originally published in Italian as Storia della luce, 1939."

 Includes bibliographical references.
1.Optics. I.Title.